S0-BWT-644

5-24-73

INTERNATIONAL
PROPAGANDA
AND
COMMUNICATIONS

INTERNATIONAL PROPAGANDA AND COMMUNICATIONS

General Editor:

DR. CHRISTOPHER H. STERLING
Temple University

Editorial Advisory Board:

DR. MORRIS JANOWITZ
University of Chicago
DR. JOHN M. KITTROSS
Temple University
DR. BRUCE LANNES SMITH
Michigan State University

transmitting world news

by Francis Williams

ARNO PRESS
A New York Times Company
New York • 1972

Reprint Edition 1972 by Arno Press Inc.

Reprinted from a copy in The Princeton
University Library

International Propaganda and Communications
ISBN for complete set: 0-405-04740-1
See last pages of this volume for titles.

Manufactured in the United States of America

Library of Congress Cataloging in Publication Data

Francis-Williams, Edward Francis Williams, Baron,
    1903-1970.
    Transmitting world news.

    (International propaganda and communications)
    Bibliography: p.
    1. Telecommunication. 2. News agencies. 3. Press.
4. Foreign news. I. Title. II. Series.
[HE7700.F75 1972]   384   72-4686
ISBN 0-405-04770-3

transmitting world news

1751833

transmitting world news

a study of telecommunications and the press

by Francis Williams

illustrations by the Nederlandse
Stichting voor Statistick
(Netherlands Statistical Foundation)

Unesco, Paris

Published in 1953 by the United Nations Educational, Scientific and Cultural Organization,
19 avenue Kléber, Paris-16ᵉ. Printed by Berger-Levrault   (Printed in France) M.C. 52 D. 14. A.

the author

Francis Williams is well known as a journalist, author and broadcaster. Prior to World War II he was editor of the London *Daily Herald*, which under his direction was the first newspaper in the world to reach a circulation of 2,000,000 copies daily. During the war he was Controller of the News, Photographic and Press Censorship Divisions of the British Ministry of Information and was made a Commander of the Order of the British Empire (C.B.E.) in recognition of his services. He was also awarded the Medal of Freedom with Silver Leaves by the United States Government, in particular for his work in organizing news coverage for the "D-Day" landings in Europe.

He was Chief Press Adviser to the United Kingdom Delegation at the United Nations Conference at San Francisco in 1945 and subsequently Adviser on Public Relations to the British Prime Minister.

Mr. Williams was chairman of the international technical committee which advised the United Nations on the establishment of a Department of Public Information and later served as British member of the United Nations Sub-Commission on Freedom of Information and of the Press.

He has been chief United States correspondent for the London *Observer*, and at present writes regularly on current affairs for the London *News Chronicle*. He was a Governor of the British Broadcasting Corporation during 1951-52.

His books, which have been published in the United Kingdom and the United States and translated into many European languages and Japanese, include the well-known *Press, Parliament and People*, a study of the relationship between governments and newspapers.

foreword

"News," as Francis Williams observes at the outset of this study, "is not merely the concern of those professionally involved in its collection and distribution, but of all men and all nations." It has been a major concern of the United Nations and of Unesco which have sought, in co-operation with the profession, to devise ways of improving the quality and increasing the quantity of information reaching the public.

Unesco's effort has been directed mainly towards improving and extending the physical facilities for the transmission of news. The present paucity of information to and from large areas of the world is difficult to accept with complacency in an era in which science has opened up boundless opportunities for full and rapid communication.

Proposals for facilitating the transmission of press messages were considered at an International Telegraph and Telephone Conference convened in Paris in 1949 by the International Telecommunication Union. The Paris meeting made it evident that effective remedial action would require further investigation of the system of rates and priorities as well as other technical factors affecting the dispatch of press messages.

Unesco commissioned Mr. Francis Williams to undertake a comprehensive study of this complex subject. The Organization's intention in commissioning and publishing the study was to place before the public and the profession, as well as governments, the essential facts—plus the opinions and recommendations of a highly qualified observer with long practical experience in press communications. Mr. Williams' precise qualifications are set forth in a preceding page, a footnote to which is the fact that he assumes complete responsibility for the whole of the study, including the opinions and recommendations.

This report has been prepared specifically with a view to action that might be taken at the next International Telegraph and Telephone Conference, to be held in 1954 or 1955. It is hoped that the book will assist in securing the adoption at that meeting of measures designed to enhance the use of telecommunications for the free flow of information.

contents

9

news is everyone's concern

If peoples and nations are to understand each other, they must know about each other. Moreover, their news of each other must be continuous and consistent; words must move constantly over frontiers and across continents and not only at moments of difficulty, tension or international passion. Such knowledge will not, of course, of itself guarantee world peace and international order. But without it, peace and order are made more difficult in our complicated yet shrinking world. Lacking it, we live as strangers among our fellows.

The free flow of news is therefore not merely the concern of those professionally involved in its collection and distribution, but of all men and all nations. The physical means by which words may be sent across frontiers are a basic professional interest of newspaper and radio men, editors, publishers, reporters and special correspondents. But they are also in the truest sense an international interest. If these means are inadequate, if there are areas of the world in which they do not exist to the extent necessary for effective news coverage, or if they are technically retarded, slow or expensive, international understanding itself suffers.

This survey is concerned with these physical means of communication. It is thus a note on a theme as old as civilization. The struggle to destroy the distances that keep men apart goes back, in one form or another, to the beginnings of history.

Much more is involved, of course, in the ceaseless effort to communicate than the physical means by which communication is made possible: the competence, integrity and good judgment of those who collect and distribute news, the readiness of governments and peoples to allow the objective reporting of affairs within their territories and its transmission across their frontiers; their equal readiness to allow the news of the rest of the world to enter their territories from many sources and to circulate freely within them; the existence of sufficient newsprint to enable what is received to be printed and published.

But none of these can fully avail unless the physical means of communication exist—and exist to the degree and in the form that modern conditions require. It is not sufficient that the means of communication shall be adequate to allow of the rapid transmission of news between the great capitals and the most industrially-developed nations of the world. It is also necessary that there shall be the means to transmit news quickly to and from areas less advanced in their economic, industrial and political development, for news of the movement of events and of opinion in such areas is of ever-mounting importance to a true understanding of the opportunities and problems of our times. It is no less important that the populations there shall know what the rest of the world is doing and thinking.

Communications must be sufficiently developed and sufficiently cheap to allow the transmission across frontiers not merely of dramatic, exciting or tragic events, but also of the background to events. They must allow those who live far away from the great political centres of the world, yet whose lives may nevertheless be touched by what occurs in them, to see events in their true perspective and to secure, not a partial and distorted picture, but one as nearly true and whole as circumstances and human judgment will allow.

If such conditions do not exist, if communications are so difficult and so costly as to allow only the barest minimum of the most exciting, surprising or

11

dramatic news to be transmitted, then telecommunications may come to serve not as the ally but the enemy of international understanding. For to know only what is sensational may be more misleading than to know nothing.

By telecommunications, with which this survey deals, is meant, in the language of the International Telecommunication Convention (1947): "Any transmission, emission or reception of signs, signals, writing, images and sounds or intelligence of any nature by wire, radio, visual or other electromagnetic systems." Such systems and processes must often appear to the uninitiated complex and difficult. But they are the direct successors of the signal fire on the hill, the drum in the forest, the courier on the horse, the messenger in the boat, the pigeon in the air, the dispatch sent by coach or by train; of all those means, indeed, by which communities have sought, since society began, to satisfy the insistent need to communicate rapidly with each other.

It is not with telecommunications as the instrument of communication between governments, or as the conveyor of commercial and financial information between business interests, or as the carrier of personal messages between individuals, important and vital to civilization as all these uses are, that we are here concerned. It is solely with telecommunications as the servant of the press, which is itself but the servant of the public and can claim such facilities and rights as it does only in their name.

To serve these press functions, telecommunications must be of such a nature as to meet the needs of newspapers of many varying sizes, times of publications, circulations, and financial and organizational strength in territories differing immensely in their economic and political development and in the tastes, sophistication and understanding of their populations. These needs include the rapid and cheap transmission of news both within the domestic territory the newspaper serves and from outside it. This survey is particularly concerned with the services across frontiers, although those within national frontiers cannot, by the nature of the study, be entirely overlooked. News across frontiers may be distributed by the world news agencies which provide a basic intercontinental service of news to many thousands of subscribers; or it may be sent to newspapers by their own correspondents. Moreover, in a large and rapidly increasing number of instances, news in this context must be taken to include not only the written or spoken word transmitted by telephone or telegraph, by cable or radio, through direct voice transmission or by symbols such as the dot-dash of the morse code or by teleprinter, but also the rapid transmission by wire or radio of photographs with a specific or universal pictorial significance and appeal.

The newspaper delivered to the door or bought at the news-stand is, whether it be large or small, illustrated or unillustrated, written for the serious consideration of statesmen and students of affairs or for the quick information and entertainment of a mass public, a messenger from the outside world. As such it is the end product of an elaborate, complicated and highly professionalized system of news-gathering in almost every phase of which telecommunications are of the highest importance.

If he is to serve his readers, the editor of a newspaper must be able to satisfy himself that whatever happens at home or abroad of interest or concern to his public will be made known to him with the minimum of delay. His function, to quote John Delane, a noted editor of the London *Times*, is "to obtain the earliest and most correct intelligence of the events of the time, and instantly, by disclosing them, to make them the common property of the nation". He is an instrument of the public interest, serving a cause which, by its very nature, "lives," to quote Delane once more, "by disclosures... daily and for ever appealing to the enlightened force of public opinion—anticipating if possible the march of events—standing upon the breach between the present and the future and extending its survey to the horizon of the world...."

Such a survey is not, in the conditions of our time, possible without adequate telecommunications services. To newspapers of all countries there must daily flow, if they are fully to perform their public duty, the news not only of their own countries or territories of circulation, but of what is significant in the events and policies of all countries.

For much of this, newspapers depend upon the world news agencies, which in turn depend in a very special degree upon telecommunications. If they are effectively to meet their responsibilities they must have their correspondents in every part of the world. They must have offices in all the great capitals, staffed by a considerable number of full-time employees of high calibre and experienced judgment. But equally they must have their correspondents, whether full-time or part-time, so placed as to be available to report any significant development or happening in any part of the world, however remote, and able to call upon telecommunications facilities to transmit without delay what they have to say. Moreover, having received this news, the news agency must be in a position to retransmit it with all speed to every newspaper which takes its service, whether in its own country or abroad. To do so it must, once again, turn to telecommunications. If telecommunications systems are inadequate, the flow of news from the world agencies cannot be adequate. If it is much easier or much cheaper to transmit news from some countries than others, the flow of news will tend to favour these countries and to be out of balance so far as the others are concerned.

As transmitters of basic news, of those facts of current happenings which all newspapers, whatever their political complexion or readership appeal, must have if they are to do their job properly, the world news agencies have a special place in the complex pattern of world reporting. But it is no less important that telecommunications systems shall be of such a nature and shall operate at such rates as will enable as many individual newspapers as possible to employ their own correspondents in at least the most important centres of world news.

In general, world news agencies aim at a high degree of objectivity. The very nature of their calling makes it expedient for them to do so, for they must serve a vast number of newspapers differing widely in their needs and opinions. It is, however, the responsibility of newspapers not only to report the news but to interpret it. They must, so far as they are able, place what happens not only in their own countries but in others against a background which will make it comprehensible to those who have no direct knowledge of the governments or peoples concerned, and who may differ very greatly in their knowledge of or interest in international affairs. Moreover, it is not desirable that events in any country shall be seen by the rest of the world through the eyes of one agency reporter alone or even those of two or three men serving competing agencies, however objective they may seek to be. In news reporting, and still more in news interpreting, safety lies in numbers.

If it is to satisfy the demands of readers who, in a world of radioed news, are increasingly likely to know the bare facts of international events before there has been time to print them, a newspaper must be able to publish an explanation of the significance of the news along with the news itself. But to explain takes more words than simply to report. The ability to do so and thus to perform the true functions of a newspaper is thus directly affected by the cost of telecommunications. This is particularly the case so far as papers of serious functions but comparatively small circulation are concerned.

The foreign correspondent of a newspaper is the ambassador of his readers: it is primarily through his eyes that the majority of them will see the country to which he is accredited: it is the impact of his words that will for most of them determine their conception of its people and policies. If high cable and radio charges force the correspondent to reduce his words to a bare minimum, or make it economically impossible for him to report and interpret anything other than the most dramatic or startling of events, the picture they receive will be distorted.

Nor is it any longer possible, as it once was, for a newspaper adequately to serve its readers simply by posting one or two correspondents to the principal political centres of the world. Few newspapers had resident correspondents in Korea prior to June 1950 and not many had even sent special correspondents to that country for a temporary visit. Yet the nature, policies and ambitions of the governments of North and South Korea were shortly to prove of crucial importance to millions of ordinary people throughout the world.

Many factors will always, of course, determine the number of foreign correspondents any newspaper will have abroad at any one time and the places to which it will send them. No newspaper can afford to have correspondents everywhere on the chance that events in some remote country may suddenly call for the services of a skilled interpreter. But it is important that the lack, inadequacy or high cost of communications shall not provide an additional barrier to the flow of news from normally inactive but potentially important news centres, or make it economically difficult for newspapers to supplement the services of resident agency correspondents in such areas by occasional visits of their own special correspondents.

Walter Gifford, former president of the American Telephone and Telegraph Company, who served as United States Ambassador to London during 1950-52, once said that he considered the function of his company was to make it as quick and easy for a man in New York to talk to someone 3,000 miles away in San Francisco, as to pass the time of day with a friend across the street. Everyone who knows the American telephone system will admit of the progress it has made towards this admirable objective. Such an ideal is more difficult internationally. Yet if men and women are to feel themselves members of one international community and learn to act as such, the ideal of swift and easy communication is a necessary one—and especially so in the field of news.

It is the purpose of this survey to show how closely the development of world reporting and world telecommunications is linked and to examine how far the communications facilities necessary to the widest possible collection and dissemination of news can yet be said to exist.

part one

1. historical background

If the issues which today concern us in the field of telecommunications and the press are to be dealt with in proper perspective, they must be seen against their historical background. The development of the modern press is inextricably linked with that of telecommunications. In its history the invention of the electric telegraph is only secondary in importance to that of the printing press; the discovery and exploitation of radio communication no less significant than the invention of the linotype and the rotary press.

It is by no accident that newspapers, even in countries already industrially advanced before the invention of the telegraph in 1837, did not begin to achieve mass readership until after that invention. The total circulation of all London daily newspapers, for example, was less than 70,000 copies (of which *The Times* accounted for some 60 per cent), when in 1850 the first Calais-Dover cable was opened—and this at a time when Britain was considerably ahead of most nations in newspaper development. Nor is it accidental that the highest degree of illiteracy in the world today is to be found in those continents, Africa with 75-85 per cent, Asia with 65-75 per cent, South America with 40-50 per cent, which are the poorest in telecommunication facilities; and that the areas of least illiteracy are those, like Europe with 5-10 per cent and North America with 10-15 per cent, where telecommunication facilities are highly developed.[1]

Historically, the development of newspaper readership has marched in step with that of telecommunications. Not only did the advance in telegraphic communication enable the press, for the first time, to present world events with a speed and urgency which excited the interest and imagination of readers to whom the more considered and contemplative mailed dispatches of an earlier era (which often necessarily dealt with events days, weeks or even months past) had made little appeal; it also helped to precipitate a revolution in the style in which newspapers were written. Mowbray Morris, manager of *The Times*, was one of the most implacable of the early opponents of the press telegram, partly, no doubt, out of a natural conservatism, partly because he recognized the challenge the new system was likely to offer to *The Times* by making more readily available to others the sources of foreign news which were so large a part of that newspaper's early strength. Yet he found it necessary in 1869 to write to T. A. Trollope, an elder brother of the novelist and an occasional *Times* correspondent in Italy, warning him that "the telegraph has superseded the news letter and has rendered necessary a different style and treatment of public subjects".

The history of international communications and the press is inalienably linked. The development of each has consistently affected the other and must continue to do so if the purpose of public enlightenment is to be served. From the days of the first telegraphic developments, news has followed the cable, and then, as the public appetite for news has increased with its supply, the demand for quicker and cheaper transmission of foreign intelligence has in its turn stimulated fresh telegraphic, cable, and in modern times, radio developments.

It is this continuous and ever potent interaction of one on the other, this essential partnership between them which, even when it has been least

---

[1] Illiteracy in population 10 years of age and over; estimate based on national illiteracy reports compiled by the Statistics Division, Unesco.

acknowledged or actually resisted, has been to the good of both, that must be remembered in any consideration of press communications at the present time. Its moral cannot be set aside without danger by governments, telecommunications systems or the press itself. The two cannot develop apart.

Although, from one aspect, the press may appear to those concerned with the management of telecommunications systems as merely one client—and not always the most profitable—among many, it rightly stands in a special position to them. It is the servant of a public interest to which telecommunication services must themselves also give pre-eminence if they are to carry out their historical purpose and accept the obligations imposed upon them by their central technological position in advancing world civilization. The importunities of the press may sometimes seem, to those concerned with telecommunications, to be wearisome, or to be lacking in understanding of the economic and technical problems which are their daily companions. They should, however, remember that these importunities have again and again in the past provided a stimulus to developments which, without them, might have been long postponed or never brought to fruition.

Nor can the press, for its part, ignore without loss the essential nature of this partnership to which it is historically committed. Telecommunication systems are more than a technical servant of the press, one among many of the material aids upon which it relies. They serve national and international interests overlapping, but not entirely coinciding with, those of the press itself and have their own contract with civilization to which they must keep. If the march of the press has affected the development of telecommunications, that of telecommunications has affected the progress and nature of the press not only directly, but also indirectly, by its influence upon the social pattern and intellectual climate of the modern world. The partnership between them is not and cannot be either an exclusive one, nor one in which the needs of one party take, as of right, predominance over the other. Their alliance has been and will continue to be fruitful to the extent that they recognize each other's requirements and responsibilities, and the obligation upon both to work in partnership.

This acceptance of the special relationship which links them did not come easily, nor is it even now in every sense complete. It developed slowly, and it is against the historical background of that development that the future course of press and telecommunications relationship can be most durably established at this time and most fruitfully charted for the future.

This historical backward glance is especially important in considering the relationship between telecommunications systems and the world news agencies which, for economic reasons if no others, are bound to be in the future, as they have increasingly been throughout the past hundred years, the main transmitters of news across the world, even though it is desirable that material conditions should be established which will enable an increase in the flow of news from individual correspondents to individual newspapers.

The development of the first of the world news agencies, Havas of Paris, Wolff of Berlin and Reuters of London, parallels that of telegraphic communication. The senior among them, Havas, was established in 1835, only two years before the invention of the first "galvanic-magnetic telegraph", as an agency for collecting and translating extracts from the principal European newspapers it received by post, and distributing them to the Paris press. Five years later, by 1840, Charles Havas had his own correspondents in most European capitals and had established a pigeon post to distribute news to papers in Brussels and London as well as Paris.

But the telegraph, "whose electric fluid" (to quote an announcement by the British Great Western Railway) "travels at the rate of 200,000 miles per second", was soon to put the pigeon post out of existence. At the same time it was enormously to increase the scope of Havas itself and make possible the rise of competitors to, and collaborators with it, in what was to prove one of the most characteristic features of the age—the development of the world news agencies.

By the means of this "interesting and most extraordinary apparatus," claimed the Great Western Railway, "dispatches can be instantaneously sent to and from with the most confiding secrecy". Those aware of the public interest in what was happening at home and abroad were quick to see the possibilities latent in that fact.

The second of the original trio of European news agencies, Wolff's Bureau of Berlin, was established as a direct consequence of the opening to the public of the Prussian State telegraph line from Berlin to Aachen in October 1849. The third, Reuters of London, came into being two years later as a direct result of the laying of the first successful submarine cable between Dover and Calais by Thomas Russels Crampton. Julius Reuter had then, as throughout his remarkable career, an instinct for "following the cable". It was by trusting this instinct that he achieved worldwide success. Thus began the new era that was eventually to put a girdle round the earth in much less than the 40 minutes of Puck's boast, and with it went a development of international reporting that was before long completely to transform ordinary men and women's understanding of the world they lived in.

This remarkable new development of communications was not at first welcomed without reserve by established newspapers. Nor did they and the new telegraphic enterprises realize at once how important they were to become to each other.

"I do not confide much in the telegraph and I would it had never been invented," wrote Mowbray Morris, manager of The Times, to that newspaper's Berlin correspondent as late as 1853. Nine years earlier, however, The Times had confessed itself indebted to this "extraordinary power" for being able to publish news of the birth of Queen Victoria's second son only 40 minutes after the happy event.

The difficulty about this "extraordinary power" was that the cost of using it was at first almost prohibitively high for ordinary news messages, and was deliberately kept so by the early telegraphic companies which tended to regard the press as a nuisance and sometimes as a rival. Rates per word were so excessive that exclusive dispatches were economically out of the question for most newspapers. In addition, the telegraphic companies endeavoured to use their control of the physical means of communication to control the sources of news also.

The battle which the newspapers and the developing world news agencies had to fight for the right to use the new system for their own purposes of independent news reporting is one of the most significant features of those early days.

The telegraph companies argued that press messages were "uneconomic"—an argument which has in various forms recurred throughout the history of the relations between telecommunication and the press. Indeed, several of them went so far as to refuse messages sent by correspondents of individual newspapers. They sought instead to use their monopoly power over communications to compel the newspapers to accept as satisfactory a service of news compiled by the telegraph companies themselves, and offered at an inclusive fee to cover collection, writing and transmission.

The larger newspapers were strong enough to resist this pressure. Their reasons for so doing, reasons which still have force as part of the permanent case for rapid and cheap world communications, even though the original attempt of the telegraph companies to control news sources has long since ended, were well expressed by Mowbray Morris in replying to such a proposal from the Berlin Telegraphische Anstalt (Telegraph Agency).

It was, he said, the first duty of those responsible for conducting a newspaper "to obtain authentic intelligence from every quarter of the globe... and for that purpose they retain correspondents whose duty it is to supply them with such intelligence. These gentlemen are of course responsible for the information they give and it is this responsibility which constitutes the chief security of their employers." The system proposed by the Berlin Telegraphische Anstalt would, he declared, "prove exactly the reverse of this and would substitute for the individual responsibility of a gentleman specially retained to serve a particular journal the absence of all responsibility necessarily implied by the very constitution of the institution in question". He added—and across the years one must applaud him—"We would much rather remain in ignorance of information conveyed in such a manner."

Not all newspapers, either on the Continent or in the United Kingdom, where the commercially strongest newspapers were in general centred at this time, had the resources to reject this attempt so summarily. The British provincial newspapers, for example, were forced for a considerable period to depend for almost their entire world and national news upon what the telegraph companies themselves were prepared to supply. It was the revolt of the press against this news monopoly which led to the ending of private ownership of telegraph systems in the United Kingdom and the compulsory taking over of the companies by the Post Office.

In 1868, under the leadership of T. E. Taylor of the *Manchester Guardian*, the Association of Proprietors of Daily Provincial Newspapers launched a formidable attack on the "despotic and arbitrary management" of the telegraph companies. Declaring that the newspapers themselves were, and must be, better judges than the telegraph companies of what the public wanted, the association announced its intention of establishing a co-operative newspaper press association to collect, write, edit and circulate news. At the same time, the newspapers inspired so strong and vocal a public demand for an investigation of the management of the telegraph companies that it forced the government of the day to appoint a Select Committee of the House of Commons to inquire into the high charges and long delays on the services operated by them. The committee was also to consider whether, in view of these factors, the telegraph services should not be brought under public ownership as had already been done in Belgium and Switzerland.

A year later the British internal telegraph system was transferred to the Post Office at a capital cost of close on £8,000,000, it being laid down by Parliament that the Post Office, while operating the telegraph services, must have no part in collecting news. The business of collecting and distributing news, previously held as a monopoly by the telegraph companies, was thereupon taken over by the new organization which the provincial newspapers had established as a co-operative enterprise. This agency, known as the Press Association, sent out its first message from London to subscribing newspapers all over the country in February 1870. Thus was inaugurated, as a direct reply to the previous exclusive policy of the telegraph companies, one of the major national co-operative press agencies of the modern world.

The attempt by electric telegraph companies throughout Europe to monopolize the advantages they derived from their control of the physical means of quick news transmission was not the only obstacle to a proper relationship between the new communications system and the press. There were also many attempts by governments to exert control over communications systems for censorship purposes, or to use such control as an instrument for blackmailing newspapers which dared to be both outspoken and independent.

During the Crimean War of 1854-56, for example, the dispatches of *The Times* correspondent, W. H. Russell, dramatically revealed maladministration in the British Army, poor provisioning of troops, and shocking conditions in army hospitals. At the instigation of an embarrassed government, a veto was imposed on messages to *The Times* over the Balaclava-Bucharest telegraph line, which had been built by the English Submarine Company under a convention between the British and French governments, whose exclusive property it then became.

Just how great were the difficulties of communications at that time—difficulties enormously increased by this veto on the use of the Balaclava-Bucharest line—is shown in a letter from Russell to John Delane, *Times* editor, explaining "what a thing it would be" if he could have a steamer at his disposal. He had previously sent to Mowbray Morris, the *Times* manager, details of a telegraph route "from Varna to Bucharest and thence to Cronstadt and so on by telegraph to England (in 70 hours)". Now in his letter to Delane he went on to declare: "I really think wonders might be done if we could get a communication between this (Balaclava) and Varna. I am certain a safe telegraph might be dispatched from the Crimea and received in London in 100 hours at latest." Failing this, he argued a steamer would be of immense advantage. "Just take for example," he wrote, "the Battle of Inkerman on the 5th November. The action was over at 2 o'clock that day. At 4 I could have been steaming across to Varna and in all probability I could have reached it by 2 or 3 o'clock the following day (i.e. 6 Nov.) having had time to write a good account of the battle during the voyage. At 6 or 7 o'clock at latest the *Tartar* would be off to Bucharest...."

But if the Crimean war brought one example among many of the attempt to use the telegraph as a political weapon against those newspapers which ran foul of the officialdom of their day, it also helped to break down the old attitude of the telegraph companies to the press.

The great volume of press dispatches brought by the war convinced them at last that newspaper messages were worth encouraging. They now began to cater for correspondents' dispatches instead of charging exorbitant rates or refusing them altogether. From this moment a greater co-operation between the two began to develop.

The need for this co-operation was underlined by the march of events. Within a decade or so, the Indian Mutiny (1857), the American Civil War (1861-65), and the Franco-Prussian War (1870-71), not to mention many less dramatic but hardly less significant international developments, had all shown the need for better press communications if public opinion was to be kept informed of what was happening and have some chance of influencing policy before too late.

Charges were still high. Thus *The Times*, whose experience has been quoted because its position at that time as the most financially strong and influential paper in the world gives its relationship to the new telegraph services particular interest, spent £5,000 on telegrams in the course of its reports on the Indian Mutiny—an immense amount for those days. As a result, however, it was able to print dispatches from Russell which effectively exposed the falsity of many of the rumours of rape and mutilation that were inflaming public opinion in Britain. *The Times* provided, in the words of the *Saturday Review*, "the means of preserving English public opinion from dangerous and disgraceful error"—an early example of the part to be played by telecommunications in the education of an informed and lively public opinion.

Across the Atlantic, the telegraph was having an effect on the development of news services comparable to that in Europe. As the telegraph replaced the "Pony Express", of romantic tradition, and linked once-scattered communities in speedy communication, American newspapers rapidly expanded their news services. But, as in Europe, telegraph charges were high. In May 1848 six New York newspapers, therefore, founded the New York Associated Press to share the total expense of news brought into New York, thus creating a precedent in co-operative news gathering that was to have great importance in the future history of the press.

Telegraph lines were advancing across the world, and hard on their heels came the newspapers and news agencies to take advantage of the facilities thus offered to report news of people to people.

Commercial interest and the ever-mounting demand for world news all pointed to the need to bridge not only lands but oceans, and link the new world and the old in a closer bond of information about each other.

In 1856 Charles Bright, a British engineer who had laid the first deep-water cable between Scotland and Ireland, and Cyrus Field, noted American cable engineer, organized the Atlantic Telegraph Company. After two failures they succeeded in splicing a cable in mid-Atlantic in August 1858. Jubilant messages were exchanged between Queen Victoria and President Buchanan, and then the first news cable to cross the Atlantic was filed in London. It reported that the Indian Mutiny had been practically quelled and that the Chinese Empire was likely to be opened to trade by an agreement then being negotiated. In the first week 730 messages were transmitted; then the electric current failed and the cable went dead.

Before the formidable preparations necessary for a new attempt could be completed, the American Civil War ended work on the project for the time being. At the same time, it demonstrated strikingly how intimately the interests of the press were now bound up in rapid communications. Lacking cable facilities, the newspapers of Europe were forced to utilize every expedient of enterprise and ingenuity to cut down the enforced delay in bringing to a reading public, now increasingly accustomed to quick news reporting, the latest developments in a war which aroused the strongest interest and emotion among readers of all kinds.

Dispatches of correspondents with both the Northern and Confederate forces, for which the reading public of the whole of Europe from London, Paris and Berlin to St. Petersburg was waiting, had to be sent by mail boat.

In an attempt to reduce the time taken, fast steamers were chartered by British and French newspapers, news agencies and telegraph companies to meet the mail boats at sea, pick up American dispatches thrown overboard in sealed wooden cylinders, and race with them to the nearest port from which they could be telegraphed to the main newspaper centres. At first the interception of the mail boats took place in the Channel. But as public interest and newspaper enterprise increased, the attempt to cut down time increased likewise. The telegraph companies, seeking to take advantage of the opportunity offered by the deep-sea cable between Ireland and Scotland, chartered tenders in Cork to intercept the American mail boats off Queenstown.

Reuter, establishing a precedent of direct intervention in providing physical links necessary for news transmission which he was later to develop in more permanent forms, secretly secured permission to erect a 60-mile-long telegraph line of his own from Cork to the little port of Crookhaven, on the south-western tip of Ireland. Sailing from here, his tenders intercepted the American mail steamers far out at sea, picked up containers coated with phosphorous so that they could be found at night, and raced back to the new telegraph station. By these prodigious efforts the time between the original mailing and the publication of the dispatches of the war correspondents was cut down by a further eight hours. It was clear that an age had arrived in which speed was to become one of the dominating factors in international reporting.

The technical means to satisfy that need were still far from satisfactory. Yet they were constantly advancing. For this advance some considerable part of the credit must go to the press which, itself driven by public demand, restlessly urged upon governments and telegraph companies alike the need for spreading the network of communications wider and wider.

In the closing days of the American Civil War, the first complete overland telegraph from London to India, via Russia, Constantinople and the Persian Gulf, was completed. Previously, the long overland wire from Paris to Marseilles and then through the Mediterranean had been improved, with a resultant cut in charges of 25 per cent. But rates were still high and delays frequent. Even after 1864, when the overland route to India was completed, messages from Bombay to London normally took between 7 and 10 days and in bad weather anything from 15 to 25 days.

It was symptomatic of increasing press preoccupation with communication problems that, at this moment, the demand of newspaper subscribers for quicker news forced one of the three original European news agencies, Reuters, to enter the communications field on an international scale for itself. Wolff of Berlin was already financially interested in the telegraph service in his domestic market. With the completion of the overland line to India, Julius Reuter now opened negotiations with the King of Hanover. He secured a concession to land a cable on the island of Norderney in the East Frisians, off the north coast of Germany, and an agreement that the Hanoverian Government would connect this cable with new land lines which it would build to Hanover and thence to Hamburg, Bremen, and Cassel for Reuter's exclusive use. By these means Reuter could link with the overland wire to India and so establish a virtual monopoly in the quickest route from London to the East.

The Norderney cable was opened for traffic on the last day of 1866. Reuter's exclusive use of it was later modified by the political intervention of the Prussian Government and four years later he agreed to sell it to the British Government (at a profit of £573,000) when the British domestic telegraphic system was taken over by the Post Office. Nevertheless the establishment of this cable and its resulting advantage helped, in the words of the Reuters official history, *Reuters Century*, to lay the foundations of the "Reuters empire". It was the beginning of an expansion which established Reuters then and for long afterwards as the primary agency for world news from Bombay to Yokohama and throughout most of the Middle East. There have been few more striking examples of the importance of command of rapid communications to a world news agency and of the extent to which telecommunication systems and the expansion of world reporting are linked together.

Meanwhile, with the ending of the American Civil War, the effort to span the Atlantic by cable was renewed. In 1865 the cable ship *Great Eastern* took a cable two-thirds of the way across to Newfoundland. Then the cable snapped. The next year a renewed attempt was successful. From Valentia, off the coast of south-west Ireland, the *Great Eastern* succeeded in laying a cable to Hearts Content, in Newfoundland. Moreover, it was able to rescue and reconnect the snapped cable of the previous year, so that now Europe and the American continent were linked by two telegraph cables. And again, after the official messages of congratulation, one of the first items to travel from continent to continent under the ocean was a press message, paid for at £2 a word, announcing the Peace of Prague that ended the war between Austria and Prussia.

Hard on the heels of the opening of the cable there came an agreement among the news agencies of the two continents. Reuters and Havas jointly arranged with the New York Associated Press to supply it with world news. Wolff had bitterly resented Reuters' incursion into German territory through the Norderney cable and had countered it by establishing his own Continental Telegraph Company, with a secret agreement with the Prussian Government that all his agency's political news messages should be counted as official, and thus given priority on all State lines over press telegrams from Reuters or other press correspondents. He now refused to join the Reuters-Havas-New York AP agreement and instead made a separate arrangement with the Western Associated Press in Chicago.

Still the cable advanced and with it the spread of world reporting. In 1869 the *Great Eastern*, fresh from her Atlantic triumphs, laid the first submarine cable to India. It was extended east to Singapore and China and eventually to Japan, less than 20 years after Commodore Perry had sailed into Tokyo harbour and opened Japan to the West. And again, hard on the heels of the new cable service, came Reuters agents to open offices in Singapore, Hong Kong, Yokohama and Nagasaki.

Meanwhile, Havas, Reuters, Wolff and the New York Associated Press, taking advantage of the developing network of world communications, declared a truce in their rivalry and solemnly agreed to share the world between them. Havas was given as its "sphere of interest" France, Switzerland, Italy, Spain, Portugal, Egypt (jointly with Reuters) and Central and South America; Reuters: the British Empire, Egypt (jointly with Havas), Turkey and the Far East; Wolff: Austria, Germany, the Netherlands, the Scandinavian countries, the Balkans and Russia; New York Associated Press: the United States. Each stimulated and followed advancing communications in its own territory and each by its exploitation of communication facilities established a virtual news monopoly. These monopolies were not finally broken until after the first world war, when they fell before the implacable hostility of the Associated Press of America to any sort of monopoly in news—a hostility from which the newspapers of all countries, and also the news agencies themselves, were eventually to benefit. The AP's campaign against any kind of exclusive news agency sphere of interest was led by its general manager, Kent Cooper, and powerfully supported by the mounting power of the American press.

That lay ahead. For the time being, what determined the pattern of world news within the bounds of these agency treaties was the advance of communications. In 1873 the submarine cable from Java reached Port Darwin. Its opening was followed by the negotiation of an agreement between Reuters and the newspaper proprietors' associations of Australia and New Zealand for Reuters to provide the whole of the Australian and New Zealand press with a service of world news. The transmission cost from London to Darwin was £1 a word, with a minimum charge of £20 per message. The following year the first cable from Europe to Brazil was completed. It too was immediately celebrated by a greatly increased flow of news, and by the opening of a joint office in Rio de Janeiro by Havas and Reuters.

As one follows the history of those years of amazing communications development, the story is always the same. Wherever the telegraph or the cable goes, the news follows. The pattern of world news agency development and the expansion of the press in country after country, the increase in popular interest in news of events in foreign lands and with it the increase in public knowledge of contemporary history, all these were linked inextricably with the development in telecommunications. The international correspondent, whether in the service of a worldwide news agency or of an individual newspaper, went where the cable led him. News is not news, at any rate on a world scale, until the possibility of transmitting it quickly is brought into being.

As the cable extended and the journalist followed in its train, the great centres of world population were immediately made aware, sometimes for the first time, always more fully than before, of events in new lands. Nor was this all. With the telegraph and the cable, it became possible to bring to people in vast areas in North America, South America, Asia, and Oceania, isolated by thousands of miles of land or sea from the world's great news centres, intelligence of events which were to shape their lives no less than that of the world as a whole.

The struggles and achievements of those early years have seemed worthy of review because they so vividly exemplify the essential relationship between telecommunications and the press. The same story was to continue as the

Since this book is published in English and French, captions for illustrations appear in the two languages in order to serve in both editions.

MAJOR OCEAN CABLE SYSTEMS
PRINCIPAUX RÉSEAUX DE CABLES SOUS-MARINS

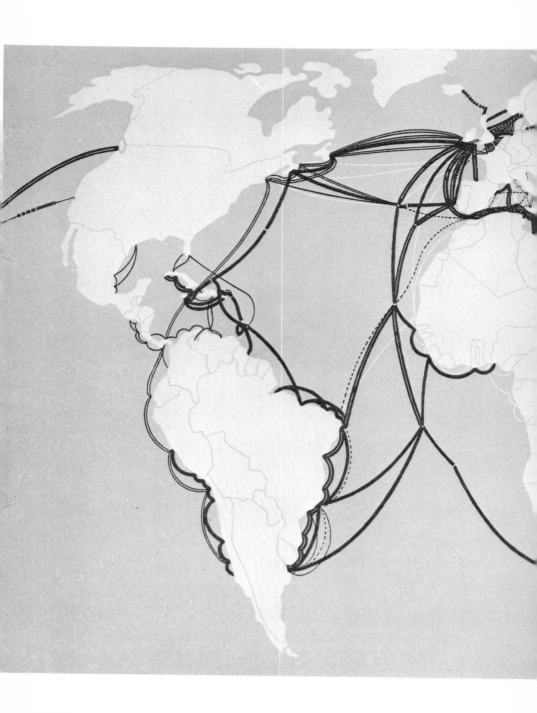

British Commonwealth
France
U.S.A.
Denmark
Italy

Double line = 2 or more cables
Double ligne = 2 ou plus de 2 câbles

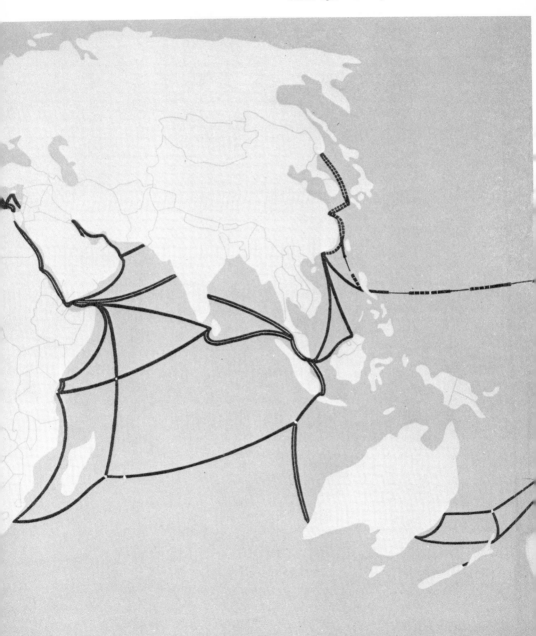

nineteeth century ended and the twentieth began. It was to receive new impetus with the sending of the first radio message in 1897, and to become ever more complex with the technical advance of the press, with the enormous growth in newspaper readership and with the development of new instruments of world information such as broadcasting.

But there was to be no change in the central theme, and no lessening in the sharpness of the moral to which all the events of those early years pointed. It is not necessary or possible for this survey to attempt a full history of these later developments. Nevertheless, the evolution of telecommunications and the press from the first invention of the "electro-magnetic telegraph" to the present time, with the consequences of that parallel advance upon newspaper style, upon the ever increasing numbers of newspaper readers, upon the education and widening interests of men and women in many lands, and upon the climate of world opinion ought some day to find its historian.

Enough has been said here, however, to show that, from the very first, the inter-relation of telecommunications and the press was fully demonstrated by events themselves, and to show also that the residual yet vastly important problems which still face governments and peoples in this field can be overcome, just as were those which faced the pioneers of world news development.

It is to the extent and nature of these present problems that we must now address ourselves.

II. *the problem in outline : needs and resources*

The close relationship between development of telecommunications and development of the press is clear from the most cursory survey of their parallel advance in the early days of telegraph and cable development—days which also mark the beginnings of the modern newspaper.

In viewing the situation today, however, and in attempting to judge how far present telecommunication services may be considered adequate for those duties of national and world reporting which are among the primary responsibilities of the press, it is necessary to break down this relationship into a number of component parts.

One must consider present telecommunications systems as they affect (a) world news agencies serving the press and radio in all parts of the world; (b) national news agencies serving the press of one country; (c) individual newspapers. Although the telecommunications needs of all these three overlap, they are not identical, and it is essential to distinguish between them.

Nor is this all. It is clearly necessary to give separate consideration to telecommunication facilities in the highly developed metropolitan areas of the world, where the volume of communications is physically adequate for all current needs—or would be so with comparatively small improvement—and those areas, accounting for a large part of the earth's surface and a very considerable segment of the world population, which still lack the physical means necessary for the adequate transmission of news.

Finally, it must be remembered that the problem of news transmission is two-directional. We must determine not only how far existing telecommunications services are adequate for the distribution of a full news service from the world's main news centres to all areas, but also how far they are adequate for a proper news service to the main news distribution centres from any area where anything of importance takes place or is likely to take place. The problem of news is always a two-way problem—that of collection and distribution.

In considering each of these aspects of the problem of telecommunications and the press, it is also necessary to bear in mind not only what would be desirable in an ideal world, but what is practical and feasible in our existing world, with all its competing demands upon economic and technical resources and available finance. It may be true that news follows the cable, or rather cable and radio. But, economically, such links cannot be brought into existence too far in advance of the forces of political, economic and social development which create the need for greater telecommunication facilities, even though such forces are themselves stimulated and helped to develop by the coming of telecommunication facilities.

It is possible and useful to chart the still empty spaces on the telecommunications map of the world. It is possible and important to consider, as we shall do later, the extent to which those areas where telecommunications are sparse or non-existent, and contact with world news is therefore small and the press undeveloped, coincide with areas where there is a high degree of illiteracy and social backwardness. But in drawing conclusions from these factors we must take into account other considerations concerning both telecommunications and the press.

It is not accidental that the parallel advance of telecommunications and the press has always been accompanied by a great increase in popular understanding and education. These two media are among the primary instruments of modern civilization. But there must be an initial demand for them to satisfy. It is too much to expect that telecommunications systems will operate without regard to the economic and social conditions in areas which are still, on an ideal assessment, underdeveloped. Telecommunication systems cannot operate in a vacuum. Their development must be related to the volume of traffic they are likely to be called upon to carry in any particular place at any particular time. Whether they are privately owned and operated as is the case in the United States, or publicly owned and operated, as in most other countries, they must be developed within a budget and cover their costs.

This dual responsibility, which is shared also by world and national news agencies, newspapers and broadcasting systems, toward the public interest on the one hand, and toward their own budgetary position on the other, must inevitably condition both the development of existing systems and the establishment of new ones. It also conditions the whole relationship between the press and telecommunications. Press demands for increased telecommunications facilities or lower transmission rates must be examined not only from the aspect of public interest, but also in relation to the same economic and budgetary factors as govern the press itself.

This warning note is sounded because the case for expanded and cheaper telecommunications can be made to appear so strong in terms of the public interest, and is in truth so strong, that it is easy to lose touch with those budgetary realities which must be fully considered in making recommendations that can be translated into reality. With the maximum understanding of and sympathy for the public interest involved, one cannot but feel that this is exactly what has happened in the case of some press demands for blanket reductions in press rates for cable and radio messages.

The Universal Declaration of Human Rights contains a clear expression of the right of all individuals to hold opinions and "to seek, receive and impart information and ideas through any media and regardless of frontiers" (Article 19). The right to receive and impart information cannot be enjoyed unless the technical facilities to transmit and receive information—and the news of current events which is one of its most important elements—exist in an adequate form. For a large proportion of the world's population, such facilities do not yet exist. In meeting this situation the press and telecommunications services share a heavy responsibility. But that responsibility is not theirs alone and cannot be borne solely by them. It is jointly shared by all those nations who proclaimed the Universal Declaration of Human Rights.

Many, although not all, of the problems which are yet to be overcome in the world's underdeveloped areas can only be met by a common international acceptance of a task which, by its nature, lies largely outside the professional and budgetary competence of the press and telecommunications systems taken by themselves, although the part they can and must play is very great.

III. *world telecommunication systems*

The complex of world telecommunications within which newspapers and news agencies must, like all other users, operate is diverse in ownership, technological advance and suitability to current world needs.

It consists of three interlinked systems: a far-flung international submarine cable system; a network of telegraph and telephone lines, comprising the domestic telegraph and telephone systems of many countries, linked to provide international channels; and an international radiocommunications system, including point-to-point radio-telegraph and radio-telephone channels and omni-directional radio transmissions, now widely used by major world news agencies for multi-address newscasts.

The relationship of these three systems to each other, the extent to which they are capable of operating as one integrated system of world telecommunications, and the degree and scope of possible future technical developments in each have all to be taken into account in reviewing the problems arising from the need for a quick, cheap and expanding means of exchanging information between all parts of the world.

The international cable system has grown up over a period of just over one hundred years.[1] During that period immense capital sums have been invested in it. Neither this investment, nor the strategical importance possessed by cables in wartime as a means of direct and secret communication, can be ignored in any assessment of the part which they will probably continue to play in world telecommunications systems in the future. But the age and efficiency of existing cables varies greatly.

[1] See map: Major ocean cable systems, facing page 24.

Some international cables laid in the 1860's are still in operation, among them the Lowestoft-Norderney cable which helped to establish the foundations of Reuter's "world news empire". Very many laid in the closing years of the nineteenth century—the peak period of cable expansion—are still in full service. The number of words they are able to carry is by modern standards low, on an average not more than 40 to 60 a minute. But the newest cables now in operation, such as the Römö-Oostmahorn cable, owned jointly by Denmark and the Netherlands, and the cable between Weybourne in the United Kingdom and Fanö in Denmark, owned by the Great Northern Telegraph Company, were both laid as recently as 1950. Numerous other cables including Anglo-French, Anglo-German and Anglo-Dutch, have been laid since 1945.

In these new cables, speeds are high and the amount of traffic that can be carried is very great. The Römö-Oostmahorn cable, for example, has 36 telephone channels, which make available about 24 times that number of telegraph channels; the Weybourne-Fanö cable has 24 voice frequency telegraph channels, compared with the single direct current telegraph channel of the great majority of the older cables.

Fourteen national administrations and 13 private telecommunication agencies are concerned in the ownership and operation of international cables; cables, that is, other than those belonging to internal systems within a particular administration's own territory, of which there are a considerable number. Two groups, the British Commonwealth and the United States, are predominant.

Since May 1948 the British Commonwealth system has been centred in the Commonwealth Telecommunications Board, which is also responsible for Commonwealth radio telecommunications. The governments of the United Kingdom, Canada, Australia, New Zealand, South Africa, India, Ceylon and Southern Rhodesia are all members of this board, and the cable (and radio) stations in their respective territories which come within the Commonwealth common-user system are controlled by national bodies in each country. Following this transfer from private to public ownership, the 155,000 nautical miles of submarine cables in the North and South Atlantic Oceans, Pacific Ocean, Indian Ocean, Mediterranean and Red Seas which make up the Cable and Wireless Ltd. system were taken over by the United Kingdom national body. Cable and Wireless, now a public corporation, has, as part of the national body, retained operational management of the cable network and of its cable stations in the United Kingdom and overseas (except for those in the territories of the other partners in the Commonwealth Telecommunications Board). The Post Office, as the other part of the United Kingdom National Body, operates the Anglo-Continental cables.

Although we are dealing at the moment only with the international cable system, it may be added for purposes of clarification that the United Kingdom National Body also owns the British international radio network because it is complementary to the cable network and closely integrated with it. This radio network is operated through the Post Office at United Kingdom radio stations, and through Cable and Wireless at overseas radio stations other than those in the territories of partner governments.

The United States international cable system, the second largest in the world, is entirely privately owned. The chief agencies are, first, Western Union, which operates 14 submarine cables measuring 30,000 nautical miles and including 8 transatlantic cables to Britain (5 of which are leased from the United Kingdom until the year 2010), 2 to the Azores and 4 to the West Indies. The other major agency is the International Telephone and Telegraph Company (IT & T), which operates in the cable field through two wholly owned affiliates. These are the Commercial Cable Company which operates 6 cables totalling 22,000 nautical miles between New York and Europe via the Azores, Novia Scotia and Newfoundland, and All American Cables and Radio Inc., which operates

5 cables of a total of 24,000 miles between the United States and South America, Central America and the West Indies. The IT & T is also a partner with Cable and Wireless and the Grande Compagnie des Télégraphes du Nord of Denmark (the Great Northern) in the Commercial Pacific Cable Company, which links San Francisco with Honolulu, Guam and Manila.

Important private cable operators outside the British and American groups include the Compagnie Française des Cables Télégraphiques, with cable services between France and the United States, Canada and the United Kingdom; the Grande Compagnie des Télégraphes du Nord, already referred to, which operates a European system linking Norway, the United Kingdom, Sweden, Denmark, France, Finland and the U.S.S.R., and a Far Eastern system with cables linking Vladivostock with Japan, and Japan with China; and the Ital-cable Servizi Cableografici, Radiotelegrafici e Radioelettrici Societa per Azione, Rome, which operates cables linking Italy, Spain and Portugal and, across the South Atlantic, Argentina, Uruguay and Brazil.

Of the 460-odd cable links, operated by national administrations and private telecommunication agencies, which are shown in the International Telecommunication Union Cable List as available to carry news about the world, only 48 of the more modern cables can provide telephone or voice frequency telegraphic circuits. The remaining 412, including all transatlantic cables, are available for direct current telegraphy only and in most cases provide only one channel.

There are no international cables capable of transmitting facsimile—i.e. the reproduction at the receiving end of the material as offered for transmission, whether written, typed, printed or drawn. Yet this method, which is now being developed in both overland wire and radio systems, seems likely to emerge as a fully automatic device, requiring little labour to operate, and bringing with it the possibility of a great increase in the volume and speed of press and other communications.

The world cable system forms, in its turn, an integral part of a still greater world network of line telegraphs. In addition to the ocean cable systems, some 700 international telegraph channels crossing national frontiers by sea or land are listed by the International Telecommunication Union. Including colonial and non-self-governing areas, they link over 130 national territories ranging from those like Haiti, Saudi Arabia, Tripolitania and Tanganyika Territory, which have only one telegraphic link with the outside world, to the United Kingdom with 33 channels, France with 23, Belgium, 19 and Italy, 18.

Of these 700 international telegraph links, only 36 are modern telex[1] channels and these for the most part connect capital cities or other large centres of population. More than 60 of the world's international telegraph channels still operate on morse, a method of transmission largely abandoned in the most highly-developed domestic telecommunications systems because of its slowness.

These international telegraph channels link internal systems which are owned and operated in most of the larger news centres by the postal and telegraphic departments of national administrations. The United States is an outstanding exception. Outside Europe, ownership of telephone channels is less frequently under public control.

From the point of view of press communications and the international need for the widest possible collection and dissemination of information concerning all peoples, the value of international telegraphic channels is affected not only by their own speed and efficiency, but by the efficiency and comprehensiveness of the national systems which feed them and which they themselves feed.

Nor for press purposes can they be separated from telephone services, which

[1] A telex service is operated on a system similar to a telephone exchange. Subscribers can connect by telephone to the exchange teleprinter.

internally furnish a supplementary but often important service of news transmission from local correspondents to offices of individual newspapers and news agencies. Internationally, fixed time telephone calls provide an important part of the press communication system between foreign correspondents and their offices, particularly for resident correspondents in large cities.

Moreover, it is the national and international telephone channels (line channels in continental land areas or radio channels in transoceanic services), which, because of their reproductive superiority, provide the main vehicles for the transmission of photographs.

The internal line systems which thus form an integral part of international telegraph and telephone communications differ, however, very widely in speed, efficiency and comprehensiveness between the most and least advanced countries in this respect.

In the United States, the Western Union Company has, in addition to operating an international cable service, developed since its incorporation in 1851 an integrated nation-wide telegraph system through the purchase, lease or stock ownership of more than 500 telegraph companies. It operates as a regulated monopoly under the provisions of the Communications Act and is subject to regulations by the Federal Communications Commission in respect of tariffs and service. Thus, although a private corporation, it in effect operates a public service comparable to the publicly administered services which are now the normal pattern in domestic telegraph services elsewhere.

Its service includes more than 1,000,000 miles of open wire and nearly 380,000 miles of wire in underground and overhead cables. The great bulk of its traffic is recorded by teleprinter or typing reperforators, but it is now also beginning to develop facsimile.

As the regulated monopoly in the field of domestic telegraph services in the United States, Western Union is—together with the Bell system, which operates the telephone service—the owner of most of the lines which make up the great teleprinter networks of the major American news agencies, networks greater in size than those in any other part of the world.

Internationally, the Western Union service links up not only with the company's own cable system, but with that of all other cable and radio-telegraph companies operating out of and into the United States, and provides the same pick-up and delivery service for such international transmissions as do nationally-operated telegraphic services elsewhere.

The American Telephone and Telegraph Company, which is the main operating unit of the Bell Telephone System, was one of the pioneers in international radio-telephone services and developed the first transatlantic radio-telephone service between New York and London in 1927. It has now an almost complete monopoly of the international wireless-telephone service from the United States and operates 57 circuits from New York, Miami and San Francisco. These link with the wire and radio-telephone services of 86 other countries with a combined total of 30,000,000 telephones.

Although the United States has developed its internal telecommunications, and particularly its internal telephone system, to a greater degree than any other country, the internal telecommunications systems of the United Kingdom and other major countries of Western Europe closely approach them. Each forms a highly integrated part of a continental and world telecommunications pattern.

The United Kingdom, both because of its geographical position and the lead in cable development it acquired in the nineteenth century, is still the biggest news collecting and distributing centre in Europe. It acts as such for much of the world news traffic of the American agencies as well as Reuters, and is served by a network of telegraph and telephone channels which link every city, town and village in the country with each other and with London.

The United Kingdom Post Office has been responsible for the internal telegraph system from 1869 on, and subsequently for the internal telephone service, as well as for telegraph services to Europe and telephone services to all parts of the world. Following the establishment of the Commonwealth Telecommunications Board, it also assumed responsibility for all external telecommunications services in the United Kingdom except for those offered by foreign telegraph companies working under licence in the country. These companies are the Danish Great Northern Telegraph Company, the French Cable Company, Western Union and the Commercial Cable Company (U.S.A.).

The British Post Office now disposes of international telegraphic transmissions over radio circuits, over the cables to Europe and over the Cable and Wireless cable system. In the year ended 31 March 1951 the total volume of overseas telegraph traffic handled was more than double the pre-war figure.

A photo-telegraph service is also available with the majority of European and Commonwealth countries, and with the United States, Argentina, Brazil, Egypt and Israel. There are, moreover, international telephone links to the whole of Europe and to 75 extra-European countries.

In France, the telegraph and telephone systems, which were badly damaged during the war, have now been reconstructed. The country is served by an excellent telecommunication network, including a very large teleprinter system, which makes possible the rapid collection and distribution of news internally and provides international telegraph and telephone links with the rest of Europe and overseas.

Although conditions vary to some extent, this pattern of a comprehensive internal system, integrated into the international telegraph and telephone system, is broadly true not only of Western Europe but of all industrially-developed countries. Technical developments in internal telecommunications are still proceeding in all these countries and will lead to further improvement in facilities. The basic needs of the press and of other users are, nevertheless, already well supplied.

This, however, by no means applies to many parts of Asia, the Middle East and most of the African continent apart from the Union of South Africa, as we shall see when we come to examine specific press needs in more detail.

The expansion of telegraphic and telephone facilities in such areas is bound to bear some relation to traffic development and cannot economically outrun it too far. Nevertheless, experience has shown that the provision of such facilities itself stimulates economic and social advance and ought properly to be regarded not simply as a response to current needs, but as an investment in the future.

The international telegraph and telephone system, widespread though it is, cannot achieve efficient worldwide coverage until the level of internal communications in the underdeveloped areas of the world is brought somewhat nearer than it now is to that of the most advanced countries. From the facts already cited, it is clear, also, that there is room for considerable improvement in some of these international telegraph channels themselves.

The third great system in the world complex of telecommunications, that of radio communications, is free from many of the problems which beset cable and line systems, such as heavy early investment in equipment which time is making obsolete. But the unlimited expansion which seemed likely in the first flush of development during the 1920's is now seen to be subject to limitations in the numbers of channels available in the radio spectrum.

Some 180 countries, including colonial and other non-self-governing territories, are now linked internationally by point-to-point radio-telegraph channels and 110 by point-to-point radio-telephone channels.[1] These radiocommunication services are operated in some instances by government administrations and in others by private companies. Despite the fact that the greatest develop-

[1] See pictograph: Point-to-point radio channels, facing page 36.

ments in the radiocommunication field are comparatively recent, the pattern of concentration follows very closely that in the older international cable and line systems.

In the field of international radiocommunication as in that of international cables, two groups, the United States and the British Commonwealth, enjoy pre-eminence. The United States international system, including circuits operated from United States overseas territories, covers a total of 158 radio-telegraph and 63 radio-telephone channels. Those operated within the common-user system of the Commonwealth Telecommunications Board total 152 radio-telegraph channels and 136 radio-telephone channels. These figures do not include the internal radio links in either group.

France and the territories of the French Union, which take third place in the international radiocommunications system, operate 148 radio-telegraph and 27 radio-telephone channels. The U.S.S.R. operates 31 international radio-telegraph channels and two radio-telephone channels, linking Moscow and New York and Moscow and London.

Although two or three great communication (and news) centres continue to dominate international radio communication as they earlier dominated international cable and line telegraph systems, the development of radio has nevertheless altered the communications map of the world in a revolutionary fashion. This is particularly true of many areas which formerly had only the scantiest of communications with the rest of the world.

The following examples illustrate the extent to which radio development has completely altered the telecommunication position of many countries, particularly in the Middle East and North and South America, and consequently their accessibility for news collection and distribution.

In the Middle East, Egypt was a few years ago linked to the rest of the world by only 3 international telegraph channels. It is now served by a total of 33 international radio-telegraph and radio-telephone links. Iran and Israel which each had 2 telegraph links now have 12 radio-telegraph and radio-telephone channels each; Lebanon 30, compared with 3; Saudi Arabia 14, compared with 1.

Among North and South American countries, Colombia now has 39 external radio-telegraph and radio-telephone links, compared with only 2 line channels; Costa Rica 39, compared with 2; the Dominican Republic 13, compared with 3; Ecuador 17, compared with 2; El Salvador 13, compared with 3; Guatemala 33, compared with 3 and Haiti, which had no line telegraphic links at all, is now served by 7 radio-telegraph and radio-telephone channels. Peru and Venezuela, which had only 3 telegraph links with the rest of the world, now have 37 radio-telegraph and radio-telephone channels each and Nicaragua 17, compared with 3.

Yet after all telecommunication channels—cable, line telegraph and telephone, radio-telegraph and radio-telephone—have been considered, there still remain many countries with only the most tenuous communication links with the outside world. Burma, for example, has only 2 channels for the external transmission of news or other information; Albania has no more than 3, Eritrea 1, Jordan 3, Liberia 2, Libya 4, and Paraguay 6.

The problem remains acute despite the fact that international point-to-point channels—cable, land line and radio—which make up the ordinary world telecommunication system, can now be supplemented through multi-directional newscasts by wide radio beams serving an area of many thousand square miles. For although these services are invaluable for carrying news to many parts of the world, they provide no answer to the other half of the problem of world news exchange—the collection of news from many diverse points.

We saw earlier in this chapter that world telecommunications depend upon a high degree of co-operation and co-ordination between three major systems.

In the same way, the difficulties and anomalies which inevitably occur within a worldwide telecommunications network made up of these varying systems, and operated by numerous national administrations and private agencies, can only be overcome by joint international action.

The machinery for such action exists in the International Telecommunication Union (ITU). It may therefore be useful to review the functions and authority of the ITU before considering in greater detail how adequately world telecommunications meet international information needs, and the extent of the problems, whether of facilities or costs, which still need to be overcome.

IV. international organization

Because telecommunications must of their nature transcend frontiers if they are to exist in anything but the most sparse and fragmentary form, the permanent machinery for international co-operation concerning them goes back much further than that in most other fields of United Nations activity, and is more comprehensive in its membership than almost any other international organization.

In its present form, as one of the Specialized Agencies associated with the United Nations, the International Telecommunication Union (ITU) derives its authority from a Convention approved at a Plenipotentiary Conference at Atlantic City in 1947.

But it traces its descent in direct and unbroken line from the International Telegraph Union, which was first formed as a European body in 1865, and it is thus one of the earliest as it has been one of the most successful organs of international collaboration.

The 1865 Convention, which established the International Telegraph Union by agreements among 20 European countries, laid down the first detailed regulations for the operation of international telegraph services. In 1885 these were widened to include regulations relating to international telephony. In 1906 a convention dealing with maritime radio-telegraphy was signed, to be revised and extended in 1912 and expanded to embrace all radio-telegraph services in 1927. The separate Telegraph and Radio-telegraph Conventions were amalgamated at Madrid in 1932 to form a single International Telecommunication Convention and it was on the entry into force of this single convention that the Union assumed its present name.

Under the Atlantic City Convention of 1947, the purposes of the ITU are clearly affirmed and re-established. They are to "maintain and extend international co-operation for the improvement and rational use of telecommunication of all kinds", to promote "the development of technical facilities and their most efficient operation..." and "to harmonize the actions of nations in the attainment of these common ends".

In particular, the Union has among its main tasks that of fostering international collaboration to establish rates at levels as low as possible consistent with an efficient service, and with the necessity of maintaining the independent financial administration of telecommunications on a sound basis.

It is also responsible for the increasingly important job of effecting allocation of the radio frequency spectrum and the registration of radio frequency assignments in order to avoid interference. The latter is a task for which the International Frequency Registration Board exists as an integral part of the Union itself.

The ITU is now worldwide in its scope and has more than 80 members. Its supreme authority is the Plenipotentiary Conference, which normally meets every five years. Below the Plenipotentiary Conference are Administrative Conferences, which meet more frequently. The permanent organs, the headquarters of which are in Geneva, consist of the Administrative Council, which normally meets once a year, the International Frequency Registration Board (IFRB), the International Telegraph Consultative Committee (CCIT), the International Telephone Consultative Committee (CCIF), and the International Radio Consultative Committee (CCIR). The ITU is thus concerned with every field of telecommunications and has vital responsibilities in all of them.

The Atlantic City Convention lays down that the raising of matters at the Plenipotentiary Conference shall not be limited to the member States themselves, but, that societies, associations or individuals may be authorized by the Plenary Assembly or by committees to present petitions or submit resolutions, provided they are countersigned or supported by the head of the delegation of the country concerned. Adequate machinery thus exists to enable press or other users to secure full international consideration of any telecommunications problem of importance, although in practice recommendations are most likely to achieve results if they are presented under the direct sponsorship of national delegations.

Despite the rivalry which characterized the relations between the first telegraph companies and the press, the ITU recognized early in its career the need for special treatment of press messages. The subject first came up at the St. Petersburg Telegraph Conference of 1875. The conference was then informed by the delegations of Belgium, France, Germany, the Netherlands, Switzerland and the United Kingdom of an arrangement between the telegraphic administrations of the United Kingdom and France for the use of lines between London and Paris by the press during the hours of night at an annual rent. It was agreed that such a system of hiring channels at reduced rents for press messages during night hours, when the telegraph lines were not fully occupied, was acceptable.

At the Budapest Telegraph Conference in 1896, France proposed the institution of a rate for ordinary press messages of 50 per cent of the general traffic rate, on the grounds that the frequent transmission of news was in the general interest. The United Kingdom, although agreeing in principle, asked that the matter should be adjourned for further consideration. It was not until 1903 that a general press rate of 50 per cent of the ordinary rate was, with a number of safeguards, accepted as desirable, although only by a small majority of 14 to 10, with 4 abstentions. Among the safeguards was a provision that press telegrams could be accepted during certain hours of the day only, and that they would be transmitted only between the hours of 6 p.m. and 9 a.m.

In proposing the resolution which marked this advance, the French delegate used words which apply no less forcefully to the problems of press communications today than they did then.

In urging a reduced press rate France was, he said, inspired by the liberal tradition which had always animated her people. She believed that a special rate for press messages was justified by the great importance the press had acquired, and by the benefit such a rate would bring in the education of opinion and the diffusion of ideas, by enabling the press to discharge its high mission more effectively.

34   Since that date, press telegrams have been under frequent review at ITU

conferences and the revised Telegraph and Telephone Regulations adopted by the International Telegraph and Telephone Conference at Paris in 1949 contain five articles dealing with them. These articles define press telegrams, regulate their handling and lay down that the terminal and transit rates applicable to ordinary press telegrams shall be 50 per cent of the ordinary telegram rate in the European system and $33^1/_3$ per cent in other relations, and that for urgent press telegrams the same rate per word shall be charged as that for ordinary telegrams over the same route. None of the provisions for press telegrams except those concerning the acceptance of such telegrams in transit are, however, obligatory.

The ITU is, of course, primarily concerned with regulating the handling of international telecommunications traffic as a whole, and not specifically with press traffic. Although recommendations regarding news material, as stated earlier, could be brought before the Plenipotentiary Conference by newspaper associations if they thought fit, there is no special organ or committee of the ITU which has any responsibility for keeping press telecommunications problems regularly under review and making recommendations regarding them.

So far as international traffic as a whole is concerned, and press traffic, of course, is part of it, the international telegraph regulations which have developed under the guidance and control of the ITU require administrations to provide sufficient direct channels and to maintain them in a state of technical efficiency. These regulations also lay down the method of computing international telegraphic charges. These charges are made up of the terminal rates of the administrations of origin or destination and of the transit rates of intermediate administrations and private operating agencies.

Maximum terminal and transit rates have for some time been prescribed for European countries, where international co-ordination in telecommunications through the ITU has existed much longer than elsewhere. Shortly before the war, at the Cairo Telegraph and Telephone Conference of 1938, an attempt was made to secure a much wider agreement, establishing unified rates per word for code and plain language telegrams throughout the Union. It was, however, found impossible to reach agreement on unified rates for telegrams to or from countries outside Europe, and it was not until the Paris Telegraph and Telephone Conference of 1949 that unified rates for code and plain language telegrams were made applicable to the whole world.

The monetary unit used in the composition of international tariffs subject to the international telegraphic and telephone regulations, and for international accounting between telecommunication administrations, is defined as "the gold franc of 100 centimes of a weight 10/31 of a gramme and of a fineness of 0.900." This unit of accounting was originally established when the gold standard was in general operation. It has now become a nominal unit only and, in the view of several countries, contributes to many anomalies; for the price fixed for 10/31 of a gramme of fine gold in one national currency may, in the present complex of managed currencies and controlled exchange rates, bear only a purely arbitrary relationship to its price in another. This is, however, a problem which is tied up with the entire present-day anarchy of international exchange rates. Its solution depends on the general problem of international exchange stabilization and cannot be found by the ITU alone.

Apart from its functions as a body for co-ordinating and regulating international telecommunications traffic, the ITU has played a prominent part in promoting and assisting technical development in every branch of telecommunications and may well play an even bigger one in the future.

Speedy and accurate international transmission depends upon the existence of telecommunications services of a comparable level of efficiency in all transit countries. The members of the ITU are required in this matter to pay due regard to the recommendations of the International Telegraph Consultative Committee (CCIT).

In international telephony, where for effective service the quality of transmission must be simultaneously satisfactory over every section of the circuit the ITU has played an even more important part through the International Telephone Consultative Committee (CCIF) and is continuing to do so. The CCIF has the responsibility for ensuring the existence of the common standards necessary for successful international operation and it conducts a great deal of important technical research into this and kindred matters at its own laboratory. It has been, moreover, the prime agent in the development of a successful switching programme for European telephone services, aimed at permitting them to take their places effectively in a continental network.

The three Consultative Committees of the ITU, the CCIT, the CCIF, and the International Radio Consultative Committee (CCIR), are now engaged, on the instruction of the Administrative Council at its June 1952 meeting, upon a programme which may prove of the first importance in developing communications in the Middle East and Southern Asia—an improvement which could be of great value in helping the flow of world news.

This is the preparation of a complete scheme "for connecting countries in the Middle East and Southern Asia with the network of major international telecommunication lines in Europe and the Mediterranean Basin, by metallic lines or by radio relay links". The aim of this study, which is being carried out under the direction of the CCIF's Joint Committee for the General Switching Programme, is to devise a plan which would initiate a new era of telecommunications development in these areas by combining new and existing systems into an international network capable of meeting the needs of expanded telegraph and telephone services and of civil aviation services, meteorology and broadcast programme relays.

At the same time, the ITU is collaborating with the United Nations Technical Assistance Administration in recruiting experts in various telecommunication fields for work under the United Nations Expanded Programme of Technical Assistance. Plans are being prepared under this programme for the reorganization and improvement of existing telecommunication systems or the development of new telecommunication networks—wire, broadcasting, point-to-point or mobile station radio—in underdeveloped areas needing aid.

It may here be noted that the ITU's most recent Plenipotentiary Conference, meeting at Buenos Aires in October-December 1952, adopted a formal recommendation to its Members and Associate Members expressing awareness of "the noble principle that news should be freely transmitted" and urging them to "facilitate the unrestricted transmission of news by telecommunication services".

From this brief outline of the history, authority and scope of the International Telecommunication Union, it will be clear that in this field at least, lack of necessary machinery to promote international co-operation and co-ordination does not offer any obstacle to advance. The ITU is sufficiently established and has a sufficiently long record of practical achievement to be able to provide all facilities necessary for careful consideration by telecommunication administrations and agencies of all those press communication problems which lie within the orbit of the public interest.

The international machinery exists. It is important that it should be so used as to ensure that no removable obstacles are allowed to hinder the maximum exchange of news and information between the world's peoples.

In that exchange three press groups are of cardinal importance: the world news agencies, the national agencies and the daily and Sunday newspapers, varying enormously in size, influence and financial resources from area to area. No true solution to the problem of promoting understanding between peoples can be found unless the telecommunication needs of each press group are taken into full account.

part two

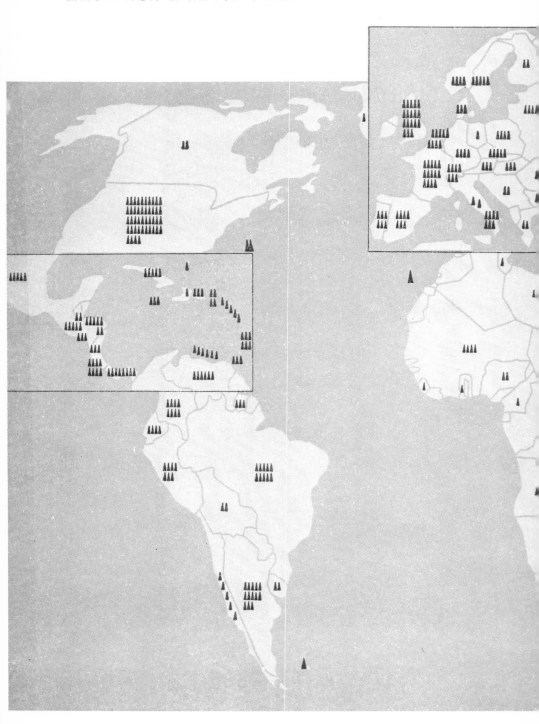

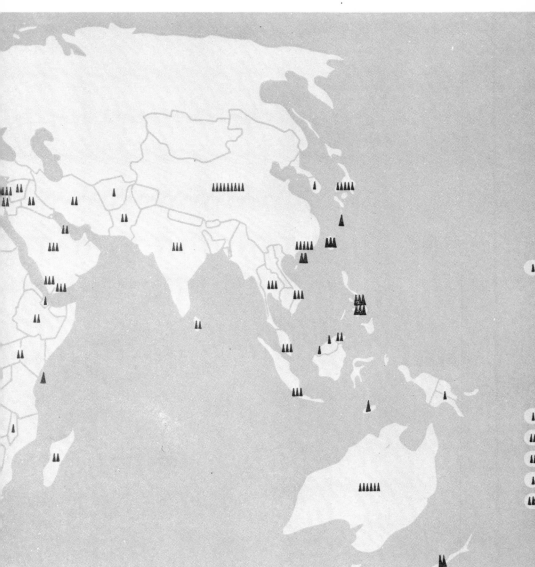

v. world news agencies

There are some 75 sizeable general news agencies in the world. Of these, 69 are mainly concerned with the distribution of news in their own national territories and 6 deal largely with its transmission across frontiers. The latter, concentrated in four national groups, are:

British Commonwealth: Reuters.

France: Agence France-Presse (AFP).

Union of Soviet Socialist Republics: Tass.

United States of America: the Associated Press (AP); United Press Associa tions (UP); International News Service (INS).

These are comprehensive world agencies offering extensive world news coverage and large-scale news distribution to subscribers in many countries.

Reuters, which in 1851 began operations in London as the private enterprise of Julius Reuter, later became a limited liability company. In 1925 this company was bought by the Press Association, a co-operative news agency owned by British provincial dailies and weeklies. In 1941 half the Press Association's interest was taken over by the Newspaper Proprietors' Association, representing London newspapers. Immediately following World War II, Reuters became a trust internationally owned by five British Commonwealth groups: the Press Association and the Newspaper Proprietors' Association (United Kingdom); the Australian Press Association (AAP); the New Zealand Press Association (NZPA); and the Press Trust of India (PTI). The AAP, NZPA and PTI are co-operative news agencies owned by the newspapers of their respective countries.

In the United Kingdom, Reuters works in partnership with the Press Association from their joint London headquarters. Reuters concentrates on British Commonwealth and foreign news, while the Press Association distributes home news, by teleprinter, to press subscribers in London and the provinces as well as in Ireland. Reuters sends its world service direct to London papers and, via the Press Association, to the provincial press.

Outside the United Kingdom, Reuters has exchange agreements with 34 news agencies in 32 countries, including AFP, Tass and AP. Within the British Commonwealth, it exchanges news with AAP and NZPA, which together cover South-East Asia and the South Pacific; and with PTI, the Associated Press of Pakistan, the South African Press Association and Canadian Press, which cover India, Pakistan, South Africa and Canada respectively. Reuters has representatives in all the main capitals. It maintains over 40 foreign bureaux, and a large number of correspondents.

Reuters depends upon radio transmission, leased wire services and ordinary commercial radio and cable facilities for the collection and transmission of news. It has been a pioneer in the development of the multiple address system for radio newscasts. (Because of its importance in world news transmission, this development will be considered in some detail at the end of this chapter.) As a general rule, the national news services to which Reuters subscribes are made available to its staff correspondents in the country concerned and a selection for general distribution is made by them.

Agence France-Presse acts both as a national and a world news agency. Founded in 1944, AFP is operating for the time being as a public establishment but provides its services under the same conditions as its pre-war predecessor, the Agence Havas, which began operations in 1835. The French Government appoints the agency's director-general, leaving the other posts to be filled by him. A bill being considered by the National Assembly aims at giving AFP an independent status. AFP's revenues are derived from subscription payments and, provisionally, from credits voted annually by the National Assembly.

The agency maintains 15 offices in France and 63 bureaux abroad. Its teleprinter network links the central Paris office with the dailies of Paris and the provinces and with eight other European capitals. Transmission is mainly by wire but in part by radio. AFP is now using experimentally a new type of radio printer known as the Coquelet system, which is expected to show considerable advantages. In addition to its European network, AFP has radio-teleprinter links with the United States, Canada, South America, Mexico and North Africa and uses transmission in Morse for sending news to Central and Eastern Europe, the Middle and Far East and South Africa.

Some 40 per cent of AFP's daily distribution service is world news. Apart from serving France and the French territories, the agency maintains exchange agreements with 30 agencies in 27 countries, including AP, Reuters and Tass.

Tass, like AFP and the three American agencies AP, INS and UP, acts as a national as well as a world agency. Established in 1918, it is State-controlled and is the sole agency for collecting and transmitting news within the U.S.S.R. Tass maintains bureaux or correspondents throughout the country and representatives in the world's chief news centres. It has arrangements, sometimes informal, to exchange domestic news with several other world agencies such as Reuters, AFP, AP and UP.

Tass is the main source of news for national agencies in countries politically associated with the U.S.S.R., such as the Hsin-Hua News Agency (New China News Agency—NCNA), Allgemeine Deutsche Nachrichtendienst (ADN—Democratic Republic of Germany), Ceska Tiskova Kancelar (Prague), Agence Télégraphique Albanaise (Tirana), Magyar Tavirati Iroda (Budapest), Bulgarski Telegrafitscheka Agentzia (Sofia), Agentie de Informatii Telegrafice (Bucarest) and Polska Agencja Prasowa (Warsaw). It now has exclusive rights for world news distribution, through NCNA, in mainland China. In addition, Tass provides news to agencies in countries outside the Communist group, such as Pars (Iran), Kantorberita Antara (Indonesia), Kyodo News Service (Japan) and Bakhtar (Afghanistan).

In collecting world news, Tass depends mainly on ordinary commercial telecommunication channels. In news distribution, it relies largely on radio transmissions supplemented by teleprinter links in certain areas. Tass makes regular newscasts which are monitored by most of the agencies which have contracts to receive them. As the official Soviet news agency, it is widely quoted.

Oldest of the three American agencies, the Associated Press had its beginning in 1848 with the foundation of the New York Associated Press. It was not until 1892, however, that AP took its present form as a nationwide co-operative agency owned by American newspaper members. Within the United States AP maintains 34 principal bureaux and 67 other correspondent points; to distribute news it uses 350,000 miles of leased wire and an extensive wirephoto network for pictures. As with UP and INS, most AP news from Europe, Latin America and Canada is routed through its New York headquarters and from the Pacific through its San Francisco bureau. AP maintains 51 foreign bureaux and has exchange agreements with nine agencies in as many countries, including Reuters and AFP.

The greater part of AP's traffic is channelled by cable or radio from London, which handles more world news than any other telecommunications centre. (There are, apart from the extremely ample radio facilities, no less then 18 transatlantic cables.) AP's London office is linked with its European leased wire system, which covers at least 12 countries. AP, like UP, uses cables extensively. Both agencies sometimes take advantage of the British Commonwealth penny a word rate for press messages by filing news from London to Montreal, whence it is sent on to New York.

About half of AP's London-New York news, however, is sent by a radio-printer channel leased from the British Post Office. This traffic is picked up and recorded by AP's listening post at North Castle, near New York, which monitors news from many parts of the world. A similar listening post near San Francisco monitors the Pacific area.

AP transmits news abroad by leased wires and radio newscasts. Leased wires are used for Western Europe and radio for Eastern Europe, the Middle East, Asia, most of Latin America and the Pacific area.

Founded in 1907 as a privately-owned service, United Press maintains 81 bureaux within the United States, with headquarters in New York. The agency operates more than 300,000 miles of leased wires to serve its United States clients. UP maintains 68 bureaux outside the United States and Canada and has exchange agreements with 14 news agencies in 12 countries. It owns British United Press, which serves Canada and the United Kingdom. Another UP service is Ocean Press, which supplies radio news daily to some 200 ships at sea.

UP's foreign bureaux are linked by an extensive system of radio transmission and leased wires. News is sent from New York to London by radio printer and thence distributed throughout Europe by a leased teleprinter network totalling over 15,000 miles. News to the Middle East, India, and Africa is transmitted by morsecasts from London and, similarly, from New York, to Latin America. To the Far East and the Pacific, news is sent by radio, partly from New York and partly from San Francisco.

Established in 1909 as a domestic agency for the Hearst newspaper group, International News Service entered the world field around 1930 and has rapidly expanded its foreign service since 1945. It maintains 34 bureaux within the United States and 21 foreign bureaux and some 5,000 correspondents and "stringers". Its domestic leased wires total 170,000 miles. INS has exchange agreements with three agencies in as many countries.

Most of its world news collection is filed through commercial channels, radio being used far more extensively than cables. Radio-printer transmissions are beamed from New York to Europe and Central and South America, while Latin America is served by morsecasts from New York. The Pacific area is served from San Francisco by radio-printer and morsecasts.

The six agencies just described between them supply news to the great majority of the world's States and territories. According to a recent Unesco survey,[1] they in fact serve as many as 144 States and territories, which together have 2,382,397,000 inhabitants, or 98.7 per cent of the world's population. This huge total is made up as follows:

AP, UP, INS, Reuters and AFP, in combination, serve 38 States and territories in Africa, North and South America, Asia and Europe, with 977,703,000 inhabitants (40.4 per cent of the world total);

Tass serves 11 States and territories in Asia and Europe, plus the U.S.S.R. itself; the whole group has 745,396,000 inhabitants (30.9 per cent of the world total);

AP, UP, INS, Reuters, AFP and Tass, in combination, serve 7 States and territories in Asia and Europe with 199,441,000 inhabitants (8.3 per cent of the world total);

AP, UP and INS, in combination, serve 21 States and territories in North and South America, Asia and Oceania with 192,542,000 inhabitants (8.0 per cent of the world total);

[1] *News Agencies: Their Structure and Opération*, 1953.

AP, UP, INS and Reuters, in combination, serve 24 States and territories in Africa, North and South America, Asia, Europe and Oceania with
96,267,000 inhabitants (4.0 per cent of the world total);

Reuters and AFP, individually and in combination, serve 38 States and territories in Africa, North and South America, Asia, Europe and Oceania, with
124,294,000 inhabitants (5.2 per cent of the world total);

AP, UP, INS and AFP, in combination, serve 5 States and territories in
Africa, Asia and South America with 46,754,000 inhabitants (1.9 per cent
of the world total).

The remaining 54 States and territories in Africa, North America, Asia, Europe
and Oceania are served by no world agency; they have 32,578,000 inhabitants
(1.3 per cent of the world total).

International leadership in telecommunications has, as one would expect from
the moral made clear in the chapter on developments in communications,
brought with it supremacy in the collection and distribution of world news.
It may incidentally be noted that the four countries primarily concerned in
the world news agency field—United Kingdom, France, U.S.S.R., U.S.A.—
likewise enjoy supremacy in the production and distribution of news films.[1]

It can be argued that while this situation may be historically inevitable in
view of the close identity between press and telecommunications, it is open
to grave objections. At first sight its practical results would appear to be
that a large part of the world must receive its picture of international events
through British, French, Russian or American eyes.

No conceivable extension or cheapening of telecommunications services would
seem likely to alter this situation. The costs of running a world news service
are so high and include so much more than telecommunications charges, that the
necessary budget can only be carried by agencies which are firmly rooted in a
strong and highly developed domestic press, with substantial technical, professional
and financial resources behind them, and with a large and established total of
subscribers to their service—or else are bountifully aided by the State.

The further development of strong national agencies, with their own correspondents in main news centres abroad to supplement the basic news services
of the world agencies, is however both feasible and desirable. So also is the
promotion of telecommunications and other facilities to enable individual
newspapers, even when comparatively small in size, to maintain more of their
own correspondents abroad either as residents in important centres or as special
correspondents in areas of particular interest.

Nevertheless, the quality and volume of the greater portion of world news
reporting seems likely to depend for some time to come upon the six existing world
agencies. In terms of the mass collection and distribution of basic news, the
problem of telecommunications is therefore primarily one of the needs of these
six news agencies. Newspapers and radio services are, of course, also concerned,
in so far as they are interested receivers of the output of the world agencies.

Although the criticism that this "four power" supremacy in the world news
agency field is dangerous has considerable force behind it and can be advanced
as an objection to telecommunications developments that may further con-
solidate the existing position, it has not the same force in practice as in theory.
The danger inherent in such a situation cannot be entirely overlooked. But
there are a number of factors which must be taken into account in assessing
it at its true value.

As we have already seen, all six agencies have an elaborate series of
exchange and other agreements with national agencies in all parts of the world,
as well as among each other. Each of the world agencies thus draws for its
news upon the national news services of many countries, apart from its own

[1] See *Newsreels Across the World*, pp. 17-18, 24-25, Unesco, 1952.

correspondents in major centres abroad, and its world reporting derives in consequence from many varied groups and nationalities.

Moreover, unless it possesses overriding political obligations to a controlling State power—a fact of which its independent international clients can be aware and which they can take into account in their use of its dispatches—a world news agency is by the very nature of its services compelled to do everything it can to achieve impartiality in news reporting. Supplying basic news to many thousands of newspapers and radio subscribers of every race, nationality, political complexion and social and religious outlook in many countries, the independent world agencies must aim at objectivity in their reports if they are to hold their clients and avoid criticism. To claim that any world agency is always completely successful in this endeavour would be excessive. But when the agencies fail as, being human, they sometimes do, it is the responsibility and duty of their clients to call them sharply to account.

Having regard to the economic conditions of our time, the real safeguard of honest and objective reporting by world agencies lies not in their reckless multiplication, but in the widest possible dissemination of their news to the maximum number of watchful subscribers, together with the closest links with the largest possible number of national agencies. Granted the telecommunications services to make this possible, the world agencies can become—as they are already to some extent becoming—not simply the collectors and distributors of news from their own correspondents, but great clearing houses of news from many sources, upon all of which they can draw to meet the information needs of many nations.

In the course of time they will, perhaps, move still nearer to such a conception. We may see the idea of international organization expanded to that of international ownership, and the example set by Reuters within the British Commonwealth widened to embrace much larger groupings, so that the national newspaper-owned agencies which the world agencies already supply, and upon which they draw for some of their news, may in time become actual partners in them. This development would be in keeping with many of the international trends of our times.

Such conjectures concern the future. Meanwhile, it is manifestly in their relationship with this small number of world agencies that telecommunications services have primarily to be considered, in so far as facilities for the collection and distribution of a truly comprehensive service of basic world news is concerned.

Each of the world agencies is the centre of a highly organized telecommunications network dealing with both the inward and outward flow of news. Although there are some differences in method, the facilities used by each of them are closely comparable. This similarity does not apply, however, as regards employment of the multiple address system of newscasts, of which some agencies have made greater use than others. The multiple address system is historically so important in the transmission of world news and has produced so great a revolution in certain services, particularly Reuters, that its development is worth describing in some detail. But for it Reuters might hardly have survived as a world agency in the difficult economic conditions following World War I. Nor could AP and UP, which had previously operated largely as national agencies concerned in the international collection of news but not in its international dissemination, have developed as world agencies to the same degree without it.

During World War I, simultaneous transmission to large numbers of receiving points by radio-telegraph had been used for propaganda purposes by various belligerent European governments. The first news agency to utilize this service was Transocean of Germany (1915). In the immediate post-war years, the omni-directional radio stations established for this purpose were largely idle.

Beginning in 1920, Reuters had meanwhile been using radio to accelerate receipt and distribution of commercial information as part of its trade service for bankers, brokers, businessmen and commercial journals. The new service was directed by Cecil Fleetwood-May, now the agency's European manager, and was the beginning of the worldwide monitoring service which Reuters, like AP and UP, now operates.

Although the first radio message across the Atlantic had been sent by Marconi in 1901, and although the *New York Herald* and the AP had established radio services for receiving news from ships at sea and sending news to them, no news agency had sought to develop the new instrument of communication on a large scale. The subsequent development of the multiple address system provides an interesting example of how a revolution in telecommunications technique can be brought about by co-operation among a group of news agencies anxious to promote exchange agreements.

Many other European agencies besides Reuters were interested in the development of speedy commercial services. The members of the European News Agency Alliance, which had been formed after 1919 to discuss mutual problems and develop exchange agreements, therefore agreed to approach their national communication services with proposals to use the now disused omni-directional stations for the exchange of commercial and financial news between themselves. Use of the stations for this purpose soon justified itself and in 1923 the main European news agencies together developed a multi-address system of international news distribution confined at first to a tightly coded financial and commercial news service.

Reuters, whose commercial service soon became the largest in Europe and was extended to other parts of the world, next undertook development of an omni-directional radio service for worldwide general news distribution.

The original cost of using the powerful war-time transmitters had been high—so high indeed that they could only be economically justified if used for a tightly coded service such as the commercial one. At the station built by the British Post Office for the Admiralty at Rugby, and first used by Reuters when it decided to extend its commercial service beyond Europe, there was, apart from other expenses, a minimum charge of £5 in starting the transmitter up from "cold" for however brief a message. This charge had frequently to be paid for two-word messages carrying urgent information of price changes to subscribers to the commercial service.

In November 1929, however, the British Post Office offered Reuters the whole-time use of a smaller but very powerful transmitter at Leafield near Oxford. This transmitter was much more economic to operate than the Rugby station, but it soon became clear that overheads could be substantially reduced if it were used for the transmission of uncoded general news as well as the coded commercial news.

By December 1929 the first continuous service of general news was being broadcast by Reuters to Europe and by the Agence Havas to South America and the Far East. Within 10 years Reuters' commercial radio news service had been transformed into a general news service known as Globereuter. Soon 90 per cent of all news sent by Reuters was transmitted in this way.

So revolutionary has been the effect on Reuters' world news service of this and other factors that strictly valid comparisons between conditions before and after its introduction are difficult to make. Some indication of its effect can however be obtained from the fact that the total number of words telegraphed abroad by Reuters increased from 115,000 a month in 1938 to 6,200,000 a month in 1951.

The Globereuter service is now transmitted on seven radio beams whose combined spread covers the entire world, except for comparatively small areas in Central America and the Antipodes.[1] The Far East is a fringe area for

[1] See pictograph: Multiple address transmission, facing page 44.

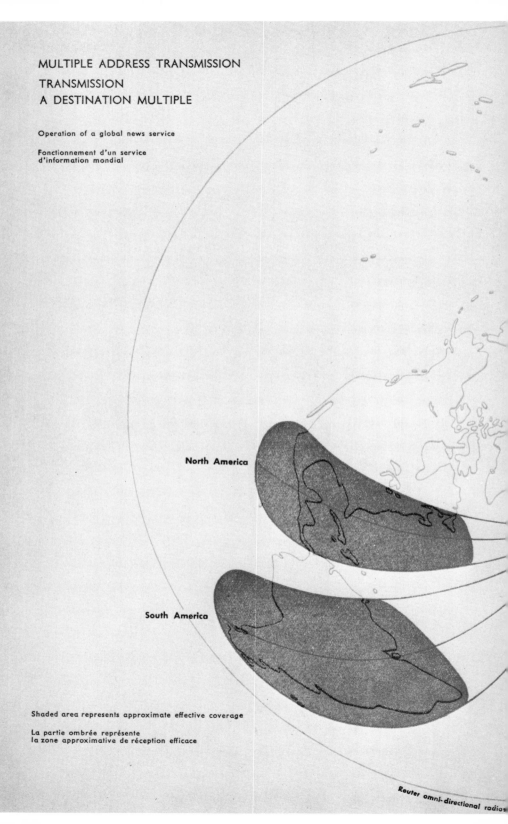

MULTIPLE ADDRESS TRANSMISSION

TRANSMISSION
A DESTINATION MULTIPLE

Operation of a global news service

Fonctionnement d'un service
d'information mondial

North America

South America

Shaded area represents approximate effective coverage

La partie ombrée représente
la zone approximative de réception efficace

Reuter omni-directional radios

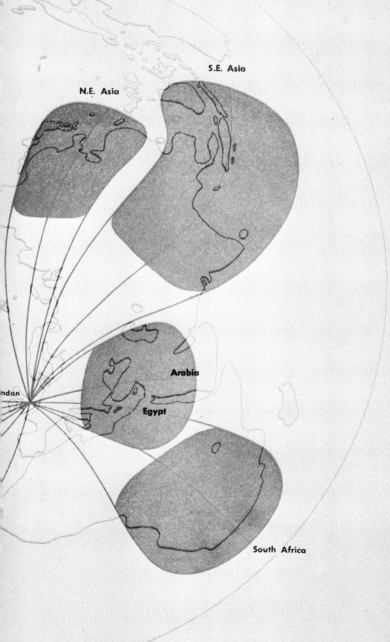

N.E. Asia

S.E. Asia

Arabia

Egypt

London

South Africa

Système omni directionnel de Reuter

direct reception, but this difficulty is being overcome by a relay in Singapore. Canada is reached by relay through land lines from the main North American receiving point in New York.

The Globereuter service, which is in fact a number of separate services going out on different wavelengths and in different directions and known as Globereuter Eastern, Globereuter African, etc., is editorially compiled by staff working at a series of editorial desks in Reuters' London office, a separate desk looking after each beam or geographical area. Transmitters are leased from the Post Office. Most of them are located at Leafield, but are actuated by direct control without relay from Reuters' regional desks in London. Telegraph operators for each beam sit beside the editors tapping out the completed copy on machines like typewriters. These machines produce a perforated paper tape which is fed into the automatic transmitter; this apparatus forms the signals which, after passing over a land line from London to Leafield, are radiated from the aerials at the wireless transmission station there.

Three systems of transmission are in use, Hellschreiber, RTT (radioteleprinter or radio-teletype) and morse. Morse is now only used on one beam, that to South America, and for a short summarized service on the African beam intended for small receiving points in East and West Africa.

Reception varies according to local conditions and regulations. In some cases it is arranged through telegraphic administrations or large private receiving stations. In others, it is picked up direct on "domestic" receivers in a small room in newspaper offices or agencies.

Originally the Reuters service to Europe was also sent out in Globereuter. This has now been replaced by a leased teleprinter network linking London with 18 European capitals. In transferring to a leased teleprinter network for Europe, Reuters followed the example of AP and UP which had earlier followed Reuters' lead in the use of multi-address newscasts, although not on so extensive a scale. The American agencies, with long experience in the use of extensive leased wire systems in their own domestic territory, appreciated the superior advantages of two-way communication offered by a teleprinter network. Being less involved in multi-directional radio transmissions than Reuters, they took an early opportunity to develop extensive European teleprinter networks soon after World War II.

Previously the development of continental teleprinter networks on the American scale had been retarded by difficulties due to national frontiers, to the gearing in of different national systems and to the necessity of reaching agreements with these over tariffs—none of which problems, of course, existed in the United States. These difficulties are now being overcome.

The further development of such continental networks, composed possibly in part of land lines and in part of radio links, may well make possible a still greater expansion of world news services, for they have two important advantages denied to the omni-directional newscast. They allow a much more direct channelling of news services to meet the needs of particular subscribers and they provide a two-way service, so that news can both be sent and received in the same channel.

We shall consider potentialities in this field in a later chapter. From the foregoing, it will be seen that the world news agencies all depend for efficient operation of their news collection and distribution services upon a complex of telecommunications facilities. These include leased wires, leased radio transmitters and the normal commercial services offered by a great number of national and international telecommunication systems. It is upon the integration of all such facilities that the successful collection and distribution of world news depends.

45

VI. *national news agencies*

As we have seen in the previous chapter, the six world news agencies greatly depend on exchange agreements with national news agencies for the maintenance of a comprehensive service and a proper balance of news. By a national agency is meant one which collects and transmits domestic news and, in some cases, distributes international news from a world agency. Government information services which distribute only official news are excluded.

Although the world agencies transmit news by radio or leased wire direct to some individual subscribers, they also depend to a considerable extent upon the redistribution of their services by national news agencies in order to secure the maximum world coverage. This dependency may increase as the pattern of international agreements further develops.

Strongly based national news agencies, served by efficient telecommunication services, are essential links in the worldwide flow of news. In areas where national agencies do not exist, world news may be received from one or more of the world agencies by one or two subscribers. But it is unlikely in many cases to pass far beyond the receiving point. Lacking agency distribution, world news cannot easily be disseminated widely enough to assure a genuine opportunity for all actual and potential newspaper readers to know and understand the significance of events abroad.

In the absence of a national news agency, world reporting of events within the domestic territory itself may, moreover, be restricted or out of balance. The world agencies can only distribute the news they receive. Their staff correspondents are chosen with care. But it is not good that events in any country should be seen by the rest of the world exclusively through the eyes of one or two men without the checks and balances—and the regular coverage of events outside the capital—which a national agency can make available.

Moreover, the absence of a national agency not only affects the international reporting of events and opinions, it tends also to make difficult an adequate exchange of news within the national area itself. For this can in most cases only be ensured by a national agency collecting news from, and supplying it to, all centres as a result of its links with individual newspapers, who in a co-operative agency will also be its members.

Furthermore, although news agencies normally come into being in response to the needs of already established newspapers, the facilities they are able to offer are among the most potent factors enabling new papers to begin. And such an extension of the press in underdeveloped areas can bring with it an increase in the general level of popular knowledge and understanding of events which is of the very greatest importance to good citizenship.

In all countries with a large newspaper readership, national agencies have now reached a high stage of organization. They have developed year by year in response to public needs, making use continuously of the new facilities offered by advancing telecommunication techniques.

But even apart from the non-self governing territories and dependencies of Africa, North and South America, Asia and Oceania, where press facilities of all kinds are, except in a few areas, woefully small, there are still some 45 States without national news agencies.[1] Some of these States, such as Egypt, Ethiopia,

[1] See pictograph: National news agencies, facing page 56.

Mexico, Cuba, Colombia, Peru, Iraq and Thailand, have large populations. Others, like Andorra and Bahrein, are small.

Illiteracy figures are available for 42 of these countries.[1] Eleven of them have an illiteracy rate of between 90 and 100 per cent, 3 a rate of between 80 and 89 per cent, 10 of 60 to 79 per cent, and 7 of 40 to 59 per cent. Only 7 of them have a rate of less than 20 per cent.

This close correlation between illiteracy and the absence of adequate press facilities, including a national news agency, is significant. It is one of the factors which has to be taken into account in any programme for promoting that freedom to receive information and to express opinions proclaimed in Article 19 of the Universal Declaration of Human Rights.

A successful national news agency cannot, however, exist without well-developed domestic telecommunication facilities. To be fully effective it must be able to rely upon a teleprinter network, whether line or radio, allowing a two-way traffic between itself and the newspapers it serves. The degree to which this is so will be best appreciated by considering the teleprinter facilities made use of in highly developed press areas.

Thus, in the United States, with three world news agencies which are also national agencies and a total of 1,773 daily newspapers having a combined daily circulation of over 54,000,000, every news centre of importance is linked by rapid teleprinter circuits.

Associated Press, which even at the beginning of the century had some 30,000 miles of leased line (a very high figure for those days, indicative of the extent to which the development of a news co-operative had already stimulated the flow of news), has now more than 10 times that amount. In the past 25 years alone its total milage of leased lines has been nearly doubled. In addition, a large wire photo network for pictures has been developed, which transmits an average of around 45 pictures a day to subscribers in almost every part of the country.

United Press operates a comparable teleprinter network with a double trunk news wire linking all main news cities as far west as the Mississippi. From this double trunk wire, news is relayed on regional circuits; the New York bureau relaying to subscribers in New England and New York State, the Atlanta bureau to subscribers in the south, and the Chicago bureau to the Pacific Coast and the north-west. The third of the United States news agencies, International News Service, has a leased wire system 170,000 miles in length.

All these press networks are leased from the Western Union Telegraph Company and other private owners and their combined cost is in the neighbourhood of $450,000 a year. Taken together, they link practically every newspaper in the country, including many of quite small and restricted circulations, to the main centres of national and world news, and have made possible a common appreciation of national and international affairs such as would otherwise have been unattainable among a population separated by immense distances and drawn from many national groups.

In the United Kingdom which, with 121 daily newspapers enjoying a combined daily sale of almost 31,000,000, has the highest newspaper readership in the world, virtually every morning, evening and Sunday newspaper published is served by the Press Association through a teleprinter network leased from the Post Office.

This teleprinter development is of comparatively recent growth. Until 1920 the Press Association distributed news in the form of press telegrams. But in that year it organized a Creed-Wheatstone automatic transmission by punched morse tape over leased wires in order to increase both the speed and regularity of its service. This system allowed a maximum speed of 140 words a minute, though average speeds were lower.

[1] National illiteracy reports compiled by the Statistics Division, Unesco.

Five years later the PA inaugurated a London teleprinter service for the national newspapers concentrated in the capital. These newspapers are now served by three channels, two for general news and one for sports news.

It was not, however, until 1949 that the present multi-channel teleprinter network covering the whole country outside London was established, with provision for switching to alternate routes in the event of any breakdown on one of the main links. Six separate teleprinter channels are available on this new system, which allows a maximum transmission at the rate of 400 words a minute and thus makes possible the speedy and effective flow of news at peak hours, however great may be the pressure of events.

In addition to the Press Association, with its almost all-embracing teleprinter network, two other agencies, the Exchange Telegraph Company and the British United Press (a wholly owned subsidiary of the United Press) also operate teleprinter services to the London papers and to many of the main provincial papers (the Ex. Tel. has some 10,000 miles of private telephone and telegraph lines in the London area alone).

Although more extensive than most, the teleprinter networks of the United States and the United Kingdom are closely paralleled by those in all highly developed newspaper areas.

Similar networks link the national agencies and newspapers of France, Belgium, Italy, Denmark, Sweden and most, although not all, Western European countries. The newspapers of Canada, Australia, New Zealand, Japan, South Africa, India, the Federal Republic of Germany and in some degree Pakistan, are also linked by teleprinter.

Not all of the major countries of the world, it is true, use teleprinter circuits to this extent; this is partly due to the great distances, partly to the destruction of communications during the war. The U.S.S.R., for example, depends to a greater degree than most of the other principal nations of the world upon radio for the internal distribution of news, either through a Hellschreiber service or through voice-casts at dictation speed. Teleprinter and ordinary telegraph services are, however, also employed. In Germany, multi-address radio transmissions, similar to those already described in the international field, are used to a considerable extent for internal news distribution.

Nevertheless, modern experience indicates that in the main the most efficient link between a national agency and its subscribing newspapers is a teleprinter network. It is, indeed, not too much to say that the stage of development reached in the collection and distribution of domestic news in any country, and the extent to which its press has the physical means to carry out fully and effectively its public responsibilities in the dissemination of domestic and world news is, in large degree, indicated by the size of the teleprinter network available.

Judged by such a standard it is clear that very many countries in the world still lack the telecommunications facilities necessary for the effective development of a national distribution of news.

Leaving aside non-self-governing territories and dependencies, the vast majority of which are still seriously lacking in adequate internal telecommunications of any kind and have only the most rudimentary of press systems, there are at least 28 States, with considerable populations, which are either without teleprinter networks or have teleprinter links so restricted in scope as to be quite inadequate to provide the basis for a full distribution of news even between the main population centres. The 28 States include eight in Europe and nine in North and South America.

Of these 28, 5 have an illiteracy rate of between 90 and 100 per cent of the population, 11 of between 60 and 79 per cent, 7 of between 40 and 59 per cent and 3 of between 20 and 39 per cent.[1]

The list of countries without teleprinter networks overlaps to some extent that

[1] National illiteracy reports compiled by the Statistics Division, Unesco.

of countries without national news agencies. But it is by no means entirely identical with it. If the two are considered together it would appear, indeed, that there are some 40 States which may be said, either because of the absence of national news agencies or the lack of adequate telecommunications facilities, to be still far behind the standard of press development which general experience indicates to be necessary to the adequate exchange of information between peoples, both nationally and internationally.

A series of comparisons between countries of somewhat similar populations will illustrate the wide diversities that are still to be found in the modern world, and how closely these diversities are related to the existence or otherwise of adequate telecommunications facilities.

Belgium, with a total population of 8,736,000,[1] of which 16 per cent is in cities of over 50,000 population, has 49 daily newspapers served by a national news service, Agence Belga, operating a nationwide teleprinter service which assures the rapid collection and dissemination of both regional and national news. Eight foreign agencies, including Reuters and AFP, have exchange agreements with Agence Belga; two other world agencies, AP and UP provide direct news services.

In contrast, Greece with a population of 7,600,000, of which 14 per cent is in cities of over 50,000 inhabitants, and with 68 daily newspapers (of which only three have a circulation of above 50,000), is without a teleprinter network. A national agency, Agence d'Athènes, exists and has exchange agreements with Reuters and AFP among others. But lack of equipment handicaps it seriously in receiving news from and sending it to the outside world. Only a few of the larger newspapers in Athens and Salonika are able to receive anything like a satisfactory service of either international or domestic news.

Consider next two Middle Eastern countries, Egypt and Iran. Egypt has a population of 20,729,000 of which 14 per cent is in cities with over 50,000 inhabitants. Iran has a population estimated at 19,140,000, of which almost the same percentage, 13 per cent, live in cities with a population of over 50,000. The geographical area of the two countries is very similar.

Egypt has 50 daily newspapers and receives national, Middle Eastern and world news from Reuters, AFP, AP, UP, and others. It also receives a specialized service of Moslem news from the Arab News Agency.

Teleprinter circuits are operated by all the principal foreign agencies and are used for distribution between the main centres.

But Iran, with 20 daily newspapers, most of which are concentrated in the capital, Teheran, and with a national news agency, the State-owned Pars Service, is without teleprinter links. As a result, the amount of international news distributed through the country generally is negligible.

In Asia, Japan with a population of 85,800,000 has 186 daily newspapers, their combined circulation being 30,218,000 copies a day. They are served by three national news agencies. These are the Kyodo News Service and Jiji Press, both of which have agreements with several foreign world and national news agencies, and Radiopress, which specializes in monitoring foreign official broadcasts. In addition, Japan is served directly by several world and national news agencies. There are several line and radio-telegraphic networks exclusively used by news agencies operating in the country, and leading news papers are equipped with teletype and telephoto facilities.

In Pakistan, which has a population of 75,842,000, the 34 daily newspapers have a combined circulation of only 120,000. Although there is now a national news agency, Associated Press of Pakistan, which has recently begun to operate a number of teleprinter lines, telecommunications facilities are extremely restricted, and such telegraphic communication as exists is slow, particularly

---

[1] Source for population figures in this chapter is Population and Vital Statistics Reports, United Nations, January 1953.

in the country districts. This lack of communication facilities for news is one of the big contributory factors to the woefully small newspaper circulation.

Next let us compare Canada, with a population of 14,430,000 distributed over a vast area of 3,603,910 square miles which presents great obstacles to the development of an efficient telecommunications system, and an Asian country, Burma, with a larger population, 18,859,000, distributed over an area less than one-fourteenth of Canada's size.

Canada has 93 general interest daily newspapers, practically all of which are members of the co-operatively owned Canadian Press. Although the first co-operative news agency was established in Canada in 1907, the absence of teleprinter facilities capable of linking the newspapers of the whole country stood in the way of a wholly national service for many years. It is significant that it was not until 1917, when a government grant of $50,000 a year (ended in 1924) made it possible to link the four main newspaper publishing areas that Canadian Press was able to develop as a truly national agency. Now it operates a system of teleprinter wires, leased from Canadian Pacific Communications, which extends right across the continent from Newfoundland in the east to British Columbia in the west, and with branches to the north and south. By means of this teleprinter circuit, Canadian Press is able to distribute news to, and collect it from, every population centre of importance in the country and can give practically every newspaper published a full service of domestic and world news. Its 250 or so teleprinters are capable, when need arises, of carrying close upon 5,000,000 words of news in a 24-hour period.

Burma, on the other hand, also has a co-operatively owned national agency, the Burma Press Syndicate, but is without teleprinter facilities. As a result, the 16 newspapers subscribing to the agency service in Rangoon have to be served by a 10,000-word roneoed daily bulletin. As for the population in the rest of the country, they receive only a slow and intermittent news service.

Next, let us make a comparison between conditions in a small but journalistically highly-developed country, New Zealand, with a population of 1,995,000, and a less developed country, Thailand, which has 19,192,000 inhabitants.

New Zealand has 43 daily newspapers served by a national news agency, the New Zealand Press Association, which has an extensive leased wire teleprinter network for distributing domestic and world news. In addition, it uses a highly-developed publicly owned telecommunications system of telegraphs, telephones, radio-telegraph and cable services. The 43 papers are located in 35 centres and daily circulation averages 368 copies per 1,000 inhabitants.

To serve a population more than nine times that of New Zealand, Thailand also has 43 daily newspapers but none of these has been able to secure a circulation much above 15,000 copies a day, and publication of all of them is concentrated in the capital, Bangkok. Daily circulation averages four per 1,000 inhabitants. There are no teleprinter circuits, and communications with the interior are so poor that no regular distribution of news is possible.

In the African continent, the Union of South Africa (12,912,000 inhabitants) has 19 daily papers served by a strong national agency, the South African Press Association (SAPA). This agency is served by an internal teleprinter network radiating from Johannesburg to sub-offices in three centres and to seven other cities, including two in Southern Rhodesia. SAPA has at its disposal extensive facilities linking South Africa with the main news centres of the world.

In the same continent, Ethiopia, with a higher population (15,000,000) and a smaller area to cover, has only one daily news bulletin and five weeklies, and is completely lacking in press teleprinter facilities.

For a comparison which is in some ways even more striking—this time on the North American continent—we may look at Haiti and Costa Rica. Haiti, with a population of 3,112,000, receives no regular news service from the outside world. It lacks not only a national news agency but any kind of teleprinter or other internal telecommunications links. Its six dailies have a 25,000 circulation, or six copies per 1,000 inhabitants.

Costa Rica, with a population of 838,000, has five daily newspapers with a circulation of 80,000 or 94 copies per 1,000 inhabitants. The country possesses teleprinter, telegraph, radio-telegraph and telephone facilities and receives world news through AP, UP and INS newscasts from New York.

Enough has been said to indicate how striking are the divergencies which still exist, despite the great developments that have taken place in telecommunications facilities and in methods of world news distribution in recent years.

The network of international telecommunications which links almost all countries in the world, plus the great advance in news distribution following the development of multi-address beam radio, now makes it practicable, as we have seen, to send world reports from the main world news centres to almost every corner of the globe. But over vast areas and in many countries, the absence of domestic telecommunications systems capable of handling the speedy and efficient collection and distribution of news by national or other agencies makes a genuine worldwide exchange of news difficult. Without such an exchange, international knowledge and understanding are gravely handicapped.

VII. *needs of the press*

"The first duty of a newspaper is to tell what happened. News is the hard core of any newspaper. All the rest of its service is merely supplemental."

These were the words of Erwin D. Canham, editor of the *Christian Science Monitor*, Boston, at a recent international press conference in Paris.

And Willi Bretscher, editor of the *Neue Zürcher Zeitung*, expressing his views on a newspaper's functions in relation to its primary task of furnishing a service of objective news, said:

"The *Neue Zürcher Zeitung* prints a large quantity of straight news as supplied by the international wire and news services. . . . But our main reliance is on our own correspondents stationed in all important and accessible news centres of the world. It is these correspondents' task to report the passing events with reference always to their background, the background which makes the particular event newsworthy in the first place and which makes its meaning clear to the reader."

Since the function of a newspaper is both "to tell what happened" and to make clear to its readers the meaning of what happened, it must always be able to supplement, with reports from its own correspondents, the basic news supplied to it by world and national agencies.

Its task in the collection of international news is different from that of the news agency. So are its telecommunication problems.

The news agency may, indeed must, rely to a very great extent upon its own leased wire and radio services for the transmission of news to its main editorial offices as well for the distribution of that news. It also, of course, depends on the ordinary commercial telecommunication services for the transmission of much of its material from correspondents in various parts of the world.

Although a newspaper may also have leased wire or radio channels between a few centres of particular importance, it depends to a very much greater extent—almost exclusively in the great majority of cases—upon the ordinary commercial cable and radio services.

The problem of telecommunications as it affects the press in relation to its own correspondents is thus primarily one of the availability and cost of facilities for press telegrams.

How large a part the cost of press telegrams plays in determining the pattern of news coverage for many newspapers may be seen from the replies to a recent survey made by the International Press Institute[1] among editors of a number of countries.

The editor of the *Sorò Amstidende* (Slagelse, Denmark) said his paper considered that a fuller reporting of United Nations proceedings would help world understanding. He had hoped to maintain a full-time staff correspondent at UN headquarters but could not do so because of prohibitive telecommunications rates.

The *Dagbreek en Sondagnuus* (South Africa) declared its experience showed that a primary necessity for a better coverage of world news was "a considerable reduction in international cable and wireless tariffs in respect of news reports from accredited correspondents to their papers". "The present tariffs", the editor declared, "have limited cable news by own correspondents to too few newspapers."

From Holland and Norway came a similar view.

The editor of *De Volkskrant*, Amsterdam, set a reduction in cable rates, particularly from Washington and New York, high among current newspaper needs, declaring that in consequence of present rates "the big newspapers are too dependent on the world agencies for the transmission of news".

The editor of the *Aftenposten*, Oslo, put the "lowest possible rates for press messages and the highest possible preference on all communication lines to secure quick transmission" among the main priorities in any means to improve world reporting.

In France, the editor of the *Parisien Libéré* declared the problem of better international reporting to be an economic one in which "the fantastically high cost of cables" plays a very important part.

And in the United Kingdom the London *Observer*, which in addition to its own usage of correspondents' reports runs a syndicated service of background foreign news to leading newspapers in many countries, reported that it would "cover many more marginal areas, such as South-East Asia, if cabling rates were lower and costs of maintaining our own correspondents less".

In India, attention was drawn to the difficulties facing underdeveloped areas from the other viewpoint—that of receiving news—as a result of high cabling costs.

"The underdeveloped countries," declared the editor of the *National Herald* Lucknow, "have no means of their own to transmit their information or even to gather it. They have to depend largely on others." He urged the need for greater equalization of communications between developed and underdeveloped countries.

---

[1] *Improvement of Information*, 1952. The International Press Institute was founded in 195 by editors from 15 countries at the suggestion of the American Society of Newspaper Editors. Headquarters have been established at Zürich with financial aid from the Ford and Rockefeller Foundations.

The object of the survey just quoted was to obtain information on measures "to promote world understanding through the free dissemination of information". The Institute, which was able to draw upon replies from 248 editors in 41 countries, concluded: "Economic difficulties preoccupy many editors in Britain, Europe, the Pacific area and South America, at times almost to the exclusion of other problems. Their main worries in this category are over the shortage and high price of newsprint and over excessive telecommunications rates. These two problems have forced many editors to sacrifice much of their foreign coverage."

The Institute went on to say that, in Europe and South America particularly, high cable rates constitute one of the most serious problems facing the press. "Editors declare it is useless for them to worry about qualitative improvements in world news reports until they can afford cable costs from such news centres as London, Tokyo, Paris and Washington."

The body of experienced newspaper opinion which holds that existing rates for press cables is one of the most formidable obstacles to the free flow of information is thus considerable.

It should nevertheless be remembered that however desirable reduction in rates may be on grounds of public interest, they must be related to the budgetary realities of the telecommunication services themselves. These services cannot be expected to subsidize one body of users, however important, at the expense of others. Nor would the majority of the world's press, which is quite rightly opposed to any form of subsidy, wish for such special assistance.

The question whether any blanket reduction in press rates is possible is closely related to the further question whether such a reduction would lead to so large an increase in press traffic as to make a reduced rate economic.

Experience of the British Commonwealth press rate of a penny a word is relevant though not wholly conclusive. This flat rate operates between all territories within the British Commonwealth, quite irrespective of the nationality of the correspondent filing a message or the ownership of the newspaper to which the message is ultimately directed.

The rate was inaugurated on 1 October 1941, largely to assure the maximum flow of information between Commonwealth countries during the war. It has, however, been continued since 1945 (although it is now in some jeopardy) and followed upon a series of earlier reductions which led to the establishment of a Commonwealth flat rate of 2¼d. a word in April 1939.[1] At this period Cable and Wireless Ltd., which had been formed in 1929 to merge the privately-owned cable service and the point-to-point radio service of the British Post Office, was carrying some 11,000,000 press words per annum within the Commonwealth.

At the Empire Press Conference organized by the Empire Press Union in June 1939, Sir Edward Wilshaw, chairman of Cable and Wireless, said that an increase of 48 per cent in press traffic would be required to make the new rate of 2¼d. a word economic. He added that in May 1939, one month after inauguration of the new rate, traffic had increased only 7.64 per cent above that of the previous January, which was an average month. Details of operating costs which he submitted are interesting as examples, although they no longer reflect the present position.

At this stage in its development the total operating revenue of Cable and Wireless was, Sir Edward declared, between £4,500,000 and £5,000,000 a year, of which Commonwealth press traffic accounted for £135,000. He further stated that, failing a compensating increase in press traffic, the 2¼d. rate would mean a reduction in Cable and Wireless receipts from press business of some £40,000 a year.

---

[1] One penny (1d.)=1.17 U.S. cents.

Sir Edward expressed the view that a further reduction to a penny rate, widely urged by the Empire Press Union, was quite impossible on economic grounds. Nevertheless it was introduced in 1941, largely on the initiative of Brendan Bracken (now Lord Bracken), who was then Minister of Information.

The results of its introduction are difficult to judge accurately. That it was accompanied by a great increase in the flow of news messages between Commonwealth centres is generally agreed. The new rate coincided, however, with an expansion in war reporting which would have taken place in any case. On the other hand, it came into effect at the same time as a great expansion in Reuters use of beam wireless for multiple address newscasts, which diverted much previous press business from Cable and Wireless channels.

Nevertheless, by 1945 Cable and Wireless was handling some 2,000,000 words of press material a week in an outward direction from London alone. When the war ended, there was naturally a decline in traffic, as the following table shows:

Year	Words outgoing from London
1946	30,101,266
1947	26,035,076
1948	30,885,432[1]
1949	26,182,407
1950	23,976,223

[1] Increase accounted for by Olympic Games in London and England-Australia cricket matches in England.

By 1951, however, the weekly outward file appeared to have stabilized itself at some 500,000 words a week (i.e. 26,000,000 for the year) and this level still rules. The weekly filing is seldom above this figure.

This is only a quarter of the peak war-time figure. Nevertheless, it shows an increase of 136 per cent over the figures given by Sir Edward Wilshaw in 1939 when he said that a 48 per cent increase in press traffic over the then level of 11,000,000 words a year was required to justify a 2¼ d. rate. It is difficult to determine how far the increase, not of 48 per cent but nearly three times that amount, compensates for the reduction to a penny rate and provides an economic basis for it. There is, in fact, considerable difference of view on this point between Cable and Wireless and the press users themselves.

That a lower rate has helped to make possible a much larger flow of news over the vast area covered does, however, appear certain.

Its full effect has, however, undoubtedly been impaired by repeated atmospheric interference with the operation of the Cable and Wireless radio beam to Australia. This beam, which normally carries a considerable proportion of press traffic, is particularly liable to ionospheric interference during which there are deep and prolonged "fades". In the autumn of 1950, an outbreak of sunspot activity led to frequent delays of 24 to 30 hours between the filing of press messages in London and their delivery in Australia, against a normal average of 30 to 50 minutes. There were comparable although less severe delays in the winter of 1951-52. This situation is now being remedied by the employment of auto-relay stations at Barbados, Colombo, Nairobi and Perth. The building of further relay stations which would cut out atmospheric difficulties completely has, however, been delayed by shortages of materials; these delays have forced press users to switch much material from ordinary to urgent rates.

Considering these limiting factors, the rise in traffic must be regarded as highly significant. There is other evidence to support the view that the penny rate has revolutionized Commonwealth press communications.

Following recommendations of the Commonwealth Telecommunications Conference, the Commonwealth cable and wireless systems in 1945 came under public ownership, each of the Commonwealth governments taking control of services in their respective countries. A Commonwealth Telecommunication's Board was established to develop the telecommunications system of the Commonwealth and Empire as a whole.

Canada is, along with various other Commonwealth countries, a member of this board. A statement made by Lionel Chevrier, Canadian Minister of Transport, at the 1950 Empire Press Conference in Quebec regarding the effect of the penny rate is therefore very relevant. It fully confirms the impression given by press file figures from London and suggests that, but for the limiting factors already mentioned, the effect of the reduced rate would have been even more striking.

"A comparison of the 1938 traffic with 1949," said Mr. Chevrier, "shows conclusively that the lowering of press rates has resulted in a great impetus to the spread of news within the Commonwealth. The volume of press traffic from Canada to all Commonwealth countries increased in 1949 over that of 1938 by 735 per cent and to Canada from all Commonwealth countries by 250 per cent."

These are remarkable figures. They leave little doubt that if worldwide reduction in press rates were possible, it would substantially increase the exchange of information between nations and enable newspapers to give their readers that wider service of international news which most editors consider necessary to the public interest.

Despite the greatly increased traffic to and from Canada that had resulted from the penny rate, Mr. Chevrier held firmly to the view that the rate was "definitely uneconomic". That view is supported by most of those responsible for operating the Commonwealth telecommunications service and there is at present strong pressure to increase the rate, perhaps to $1\frac{1}{2}$d. a word.

A detailed inquiry into the basis on which costs for press messages are calculated would throw more light on this matter. It would be of great value not only to maintenance of the Commonwealth press rate but also to possible review of press rates throughout the world.

It is clear that lower press rates can only be justified economically if the reduction in revenue per word is compensated by a large percentage increase in total traffic. But what is that percentage, in view of the argument that the penny rate remains uneconomic, despite the immense increase in traffic it has brought about?

Many specialists of the press who believe that reduced rates are desirable in the public interest and could be made economic argue that existing rates are often calculated on the basis of too great an allowance for past investments in cables. They also declare that much press traffic can be sent more cheaply by radio and the rate should therefore be fixed on that basis.

Replying to Mr. Chevrier at the conference already mentioned, Walton Cole, editor of Reuters, challenged the statement that the Commonwealth penny rate was uneconomic. Technical advances in method and equipment should, he declared, be considered. "If the press rate is uneconomic," Mr. Cole added, "it is due to the methods by which it is now being worked . . . when the point is made that the penny rate is uneconomical, that statement relates to the very large volume of investments involved."

Illustrating the point that many press rates were calculated on a basis which would not be applicable if modern methods were utilized, he reported that Reuters was radioing a telecircuit from New York to London at a rate which had proved to be well below the 4½ cents a word for commercial channel. Mr. Cole further argued that, if the Commonwealth Press Union were itse' allowed to organize a telecommunications unit to handle Commonwealth pres traffic, it could, by co-operative effort, not only maintain but actually reduc that rate.

A similar belief that the heavy investment of existing telecommunicatio systems in cables and land lines handicapped press traffic led in 1929 to th organization of Press Wireless Inc., on the initiative of a number of Unite States newspapers.

By 1936 Press Wireless, operating a telecommunication system exclusivel for press purposes, had six international radio-telegraph circuits in full us. By 1950 it had increased this number to 17, and it is still developing. I addition to its point-to-point service, it operates an extensive multiple addres beam radio system in the development of which it was a fellow pioneer wit Reuters and the British Post Office. This system has greatly expanded sin the last war.

The history of Press Wireless indicates fairly clearly that press traffic, route via radio channels, can be fully economic even when separated from the mo profitable commercial traffic. But too many conclusions cannot be draw from this fact. Over the whole field of international telecommunication radio links will doubtless play an increasingly important part in the futur But any attempt to solve the over-all problem of cheap press communicatio by an exclusive concentration on radio services would be unrealistic.

This is due, first, to the fact that heavy past investment in line systems cann be ignored; no industry can break entirely free from its past. Secondly, th development of radio services for news and other purposes is increasingl conditioned by the amount of radio spectrum space available. The spectru is already overcrowded, and allocation of frequencies between national an international claimants has become a world problem of increasing con plexity.

Each use of the radio spectrum, whether for an international news transmissi over thousands of miles or for a short-range shore-to-ship navigation signa requires the establishment of a channel in the spectrum. And although n all parts of the spectrum are either equally useful or in equal demand, ea is in sufficient demand to create allocation and assignment problems.

Moreover, the high-frequency portion of the radio spectrum (4 to 27.5 meg cycles) which is needed for medium and long-distance telecommunication is in greatest demand of all since it is also required for long-distance and tropic broadcasting.

A further complication is that possibilities of the high-frequency spectru have to be measured in terms of possible circuits rather than frequencies, sin more than one frequency assignment is, in general, required for each circui This point was stressed in an authoritative survey of telecommunicati problems—the report of the President's Communication Policy Board President Truman in March 1951.

The number of possible circuits which can be carved out of the spectru depends upon the type of circuit required (radio-telegraph, radio-telephone, et the geographical location of the terminals, the time of day, season of the yea phase of the sunspot cycle, number of circuits operated and many oth factors.

To provide a continuous 24-hour daily service of the kind required for inte national news transmissions, it may, for example, be necessary to use five more frequency assignments for a single circuit in one direction. Even wh conditions are favourable, an average of three high-frequency assignments a required daily for each one-way circuit of this kind. And as in practice su communications usually involve a two-way exchange, and therefore requi

NATIONAL NEWS AGENCIES
AGENCES NATIONALES D'INFORMATION

States with national telegraphic agencies
Pays où fonctionnent des agences
télégraphiques d'information

States without national telegraphic ne
Pays où ne fonctionnent pas d'agenc
télégraphiques d'information

NATIONAL NEWS AGENCIES
AGENCES NATIONALES D'INFORMATION

States with national telegraphic agencies
Pays où fonctionnent des agences
télégraphiques d'information

States without national telegraphic ne
Pays où ne fonctionnent pas d'agence
télégraphiques d'information

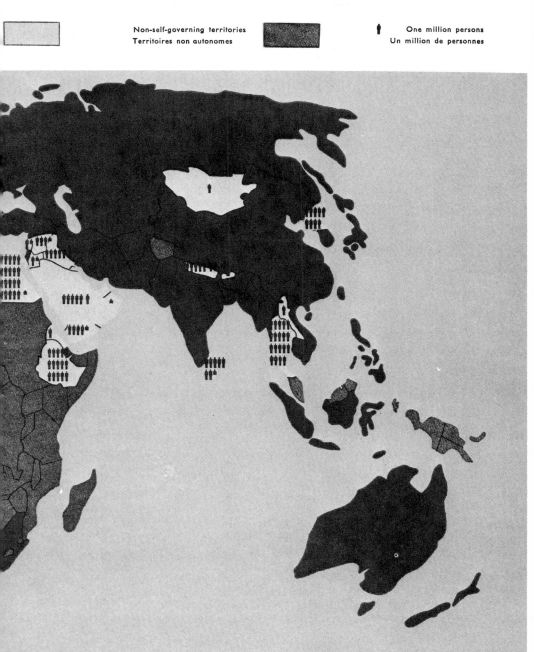

wo one-way circuits, the complement of frequency assignments is double,
,e. six.

Taking into account other technical factors which create the greatest demand
that for frequencies below 8 megacycles), the maximum possible number of
4-hour long-distance circuits is largely determined by the supply of channels
etween 6 and 8 megacycles. The pressure of demand on this portion of the
pectrum may be seen by the fact that the number of listings between 4 and
0 megacycles in the ITU frequency list increased from 1,698 in 1929 to
1,456 in 1949.

The actual number of separate and distinct channels within any particular
ortion of the spectrum is not static, and has in fact been greatly increased by
mprovements in equipment and operating techniques which have made
ossible the use of poorer grades of circuits without serious effect on the service.
Nevertheless, the possibility of considerable further improvement in this respect
s believed to be small.

More could probably be done with available frequencies by better management
f the spectrum. Continuous attention is being given to this problem by the
nternational Frequency Registration Board. It is also doubtless true, as the
President's Communications Policy Board stated, that there exists the "oppor-
unity for more effective sharing of frequencies, for more intensive use of
ndividual frequencies and for increased economy in kilocycles assigned to
ach circuit". But whatever may be achieved in this direction, the fact
emains that "contrary to the layman's opinion, the present usable spectrum
oes not offer an unlimited number of channels" (same source). Indeed, so
imited is the number of channels and so difficult is it to reconcile the expanding
lemands of sound broadcasting, television, aviation and telecommunication
ervices, that the Table of Frequency Allocations agreed at the Atlantic City
nternational Conference in 1947 is still far from being fully implemented.

Despite popular conceptions, it remains true that the land and the sea are
nuch less crowded than the air. This basic fact must always be considered
n reviewing telecommunication problems.

Even if one could ignore the economics of the telecommunication industry
nd its heavy investment in line systems and international cables, it would
herefore be impracticable to think of a solution to the problem of cheap world
ress communications exclusively in terms of radio. Nor would this be in
eeping with present technological advance.

As recently as 1946 there seemed much to justify the statement made by two
American observers, Llewellyn White and Robert D. Leigh, in a notable brief
urvey, *Peoples Speaking to Peoples*.[1] "The fact is that cables are rapidly
ecoming obsolete. . . . Cables now serve as a handicap to the development of
heap, universal service—a handicap which should be eliminated in any sound
ostwar programme."

Current technical developments affecting submarine cables and land lines have
mpaired the validity of that statement, as we shall see in a later chapter.
Indeed, many of the most important advances toward cheap and rapid transmission of news now seem likely to lie in the field of line communications.

Any sound policy for future international press communications, should therefore, be based on an integrated partnership in which radio and line systems will
play an equally vital part.

It is in the light of such a partnership that we must consider whether a standard
world press rate is achievable and desirable, and seek a remedy for the many
striking press rate anomalies now existing within the pattern of world telecommunications.

[1] *A Report on International Mass Communication from the Commission on Freedom of the Press*,
University of Chicago Press, 1946.

VIII. *the cost of transmitting news*

It will be clear from what has gone before that the principal telecommunication facilities used in the worldwide collection and distribution of news fall into four main categories.

There is, first, the transmission of messages filed at so much a word through the ordinary cable, telegraph and radio services commonly available. Upon this facility all sections of the press—the world news agencies, national agencies and individual newspapers—depend to a great degree for the means to carry out their public responsibilities.

Secondly, there is the two-way transmission of news by leased international point-to-point circuits, whether wire or radio. This facility is particularly important to the world news agencies, although it is also used in some measure by individual newspapers. It is the principal means by which national agencies and newspapers in the most highly developed areas receive their news from the world agencies.

Thirdly, there is the multiple address system of broadcasts. As a means of news distribution on a large scale, this is used primarily by the world news agencies. It is, however, no less important to their clients, whether national agencies or individual newspapers, and is particularly so in underdeveloped areas lacking adequate telecommunications facilities.

Finally, there are the internal teleprinter and other networks. These are of fundamental importance to national news agencies for the collection of national news and the domestic distribution of national and international information. Without such facilities, it is practically impossible for a national news agency to exist in a fully effective form.

How adequate for the service of the press are these four categories of telecommunication facilities at the present moment? Let us consider them in order.

The requirements by which facilities in the first category (ordinary cable, telegraph and radio services) must be judged as conveyors of news are threefold.

They should be sufficiently cheap to permit their regular use by correspondents not only of the powerful world agencies but of individual newspapers—some of them small in size and financial resources, even when highly important in influence. Rates should permit such use not only for brief reports giving the bare elements of dramatic or sensational events, but also for messages of explanation and background.

They should also be sufficiently uniform in cost not to affect the true pattern of world news exchange. Such a distortion can occur when it is relatively so much dearer to report events and background news in one area than in another that some parts of the world are left uncovered, or reported much more sparsely than would be the case if intrinsic news value were the sole standard.

Finally, they should be rapid, reliable and not subject to delays which may destroy much of the value of cabled news messages. All news messages have "deadlines"—the edition times of the newspapers for which they are intended.

There are still many international telecommunications services which fail to live up to all these requirements and a number which fail to live up to any of them.

The charges for cabling news, whether by wire or radio, vary enormously from point to point.[1] In many cases, rates appear to bear little or no relation to real

[1] See pictograph: Disparities in press rates, facing page 72.

costs, at any rate to costs in efficiently run services. In addition, the lack of uniformity is often staggering. This is frequently so, even in the case of traffic between identical points: it may cost twice as much to send news in one direction than in the other, as is shown by the following table of rates between four major news centres—London, New York, Paris and Moscow.[1]

London—New York, 2.04 cents;   New York—London, 5.54 cents;
London—Paris, 1.75 cents;   Paris—London, 2.91 cents;
London—Moscow, 3.50 cents;   Moscow—London, 8.75 cents.

These are only four outstanding examples of a disparity so common as to be accepted as a normal feature of telecommunications practice. The same situation exists, for example, in traffic between London and Switzerland. In one direction London to Switzerland, it is 1.75 cents a word. In the other direction, Switzerland to London, it is 4.37 cents a word. To give one other European example—this time in the Near East—the press rate from Greece to London is 8.75 cents a word, while from London to Greece it is 2.91 cents.

It is difficult to find a good reason for such differences of rates between identical points. Since the same systems are used whether messages are travelling in one direction or another and identically the same handling is involved, either the rate in one direction must be economically too low or that in the other direction economically too high—both cannot be right. And in the eyes of users, the existence of such disparities is bound to throw doubt upon the whole system of costing upon which international press and other rates are based.

It is vital to international understanding that there should be ample and reliable reporting of events in areas of the world which have now become, or are likely to become, the centre of important political, social and economic changes. The Middle East is such an area. But cable rates (using the term according to common press usage to cover international messages sent either by line or radio) to and from Middle Eastern countries not only vary enormously from country to country and according to which direction they are sent, but are frequently so high as to preclude anything like a full and regular service of background news. The extent to which this is so is shown in the following table.

Centre	Ordinary press rates (per word)		Urgent press rates (per word)	
	From London	To London	From London	To London
	U.S. cents	U.S. cents	U.S. cents	U.S. cents
Iran	3.50	10.50	17.50	52.20
Lebanon and Syria	3.50	9.91	10.50	29.57
Turkey	3.50	9.77	7.00	19.54
Sudan	4.37	6.27	12.83	18.96
Iraq	4.67	4.67	17.50	17.50
Egypt	2.91	3.79	9.33	11.52
Aden	1.17	1.60	7.00	6.29
Cyprus	1.17	1.17	5.83	12.39
Israel	1.17	1.17	7.00	8.75
Jordan	1.17	1.17	5.83	7.00

[1] For purposes of simplicity, all rates are here shown in U.S. cents; one British penny=1.17 cents; one French franc=0.29 cents.

Press rates to and from London have been taken because London acts as the redistribution point for the great mass of Middle Eastern news handled by the major world agencies. In addition, British newspapers are traditionally interested in Middle Eastern affairs, so that the direct flow of news between the Middle East and London is greater than between that area and any other news centre. The rates given are those ruling in August 1952. In a number of cases still higher rates are likely to come into operation in the near future. This table illustrates how very much higher in most cases is the cost of sending reports of Middle Eastern developments to London, from whence it can be widely redistributed to other parts of the world, than is that of sending news from London to the Middle East. Yet it is vitally important that there should be an adequate two-way flow of news between the Middle East and the major news centres most concerned in its worldwide distribution.

There appears to be nothing to justify this divergence even though it is, as shown earlier, paralleled elsewhere. In the case of Iraq, where basic conditions of service do not differ from those in neighbouring countries, the rate for ordinary and urgent press messages in either direction is identical. Yet in the adjoining State of Iran, the rate for outward press telegrams is three times that for inward ones. To make confusion worse confounded, it costs 1.17 cents a word *less* to send news to Iran than to Iraq, and 5.83 cents *more* to send news from Iran than from Iraq. The Iran rate, is in fact, twice as much as the Iraq one.

If this were all, it would be serious enough. But, as the table shows, the cost of filing news from the four Middle Eastern countries at the top of the list, Iran, Lebanon, Syria and Turkey, is between eight and nine times as much as the cost of filing it from neighbouring countries at the other end of the scale.

When costs differ to this extent it is impossible for them not to affect the pattern of world news coverage, particularly that of background news. Moreover, such a rate per word as that charged for press messages from Iran, for instance, makes individual coverage of Iranian news impossibly expensive for most newspapers. Taking 1950 as a representative year, it is not accidental that there were then only five full-time foreign correspondents in the whole of Iran, and only two representing individual newspapers—the London *Daily Express* and the London *Daily Telegraph.* This despite the fact that, although Iranian developments had not reached the stage of dramatic front-page news, a regular flow of background information disclosing the movement of opinion was of great importance in the light of what was to come.

High as these Middle Eastern press rates may be, they are nevertheless exceeded by rates in many other areas, as the following examples show:

Centre	Ordinary press rates (per word)		Urgent press rates (per word)	
	From London	To London	From London	To London
	U.S. cents	U.S. cents	U.S. cents	U.S. cents
Belgian Congo	4.96	14.58		44.19
Afghanistan	5.54	21.29	16.33	42.70
Brazil	7.00	15.02	21.00	64.17
Peru	7.00	14.00	22.17	63.73

In general, press rates from Central and South America to London are so high as to place the gravest obstacles in the way of adequate news coverage. This is to some extent balanced by the fact that there are, in general, fairly reasonable press rates from Central and South America to New York, the other great world news distribution centre. The rate from Peru to New York is, for example, less than a quarter of that to London and this order of difference is fairly general.

How little relation distance often has to the cost of sending news may be seen by the fact that the ordinary press rate from Venezuela to London is no less than 27.27 cents per word, whereas from the adjoining territory of British Guiana it is only 1.17 cents. When a few miles, distance on one side or another of a national frontier can alter a press rate by nearly 2,400 per cent, it is difficult to discern any logical pattern in international telecommunication charges as they relate to the vital public interest of the exchange of news between nations.

The existence of special press rates within particular groups as, for example, the penny rate in the British Commonwealth and the preferential rates between French territories only serves to underline what appears to be an almost total lack of logic in many communication charges.

The first of the following two tables gives the cost of cabling press messages from points within the British Commonwealth system to London, compared with that of cabling from exactly the same points to Paris (approximately the same distance). The second table makes a similar comparison between the cost of cabling from a number of points in the French system to the same two cities.

From	To London (per word)	To Paris (per word)
	U.S. cents	U.S. cents
Singapore	1.14	19.14
Johannesburg	1.14	13.14
Montreal	1.14	9.14
British Somaliland	1.14	4.86
West Indies (St. Lucia)	1.14	4.86

From	To London (per word)	To Paris (per word)	To London via Paris (per word)
	U.S. cents	U.S. cents	U.S. cents
Tunis	4.67	0.12	3.03
Algiers	4.67	0.12	3.03
Dakar	15.17	4.97	7.87
Rabat	8.17	3.21	6.12
Saigon	9.05	3.79	6.71

Moreover, the pattern of international press rates has developed in such a piecemeal fashion that costs between two points may vary enormously according to the way in which the message is routed. Thus, from Japan, which is the filing point for all dispatches from war correspondents with the United Nations forces in Korea (dispatches are relayed from Korea to Japan by military channels), the ordinary press rate to London is 15.47 cents a word; the urgent press rate is 56.29 cents a word. This compares with an ordinary press rate to New York of 10.50 cents, or 5.25 cents a word less, and an urgent press rate of 16.04 cents, 40.25 cents a word less. But if messages from Japan to London are routed to Hong Kong and retransmitted from that point, the ordinary press rate is only 7.58 cents a word, 7.87 cents a word cheaper than the direct rate, and 2.91 cents a word less than to New York. The urgent press rate is 26.62 cents, 29.75 cents a word less than the direct route, although still higher, by 10.50 cents a word, than to New York.

These examples plainly indicate the chaotic nature of the telecommunication press rate structure and the lack of any logical standard of comparison between one rate and another, and they could be continued almost indefinitely. This system has been produced by piecing together, over a number of years, a conglomeration of small unrelated parts. Each had a certain validity in a particular area at a particular period of technical development, but the final result is an international structure not only totally lacking in internal logic or co-ordination but, it would appear, without much relevance to the technical and other circumstances of the present time.

Certain European countries have inherited from the earlier days of the International Telegraph Union agreements on the sharing of terminal and transit charges for international messages and the fixing of maximum rates for terminal charges in the handling of such traffic. Yet even in these countries, there are great differences in actual practice. In addition, the same rates are charged for dispatches sent by direct point-to-point radio transmission, where no intermediate transit handling is involved, as for those sent by line communications, which may have to be handled by a considerable number of intermediaries, all of which must receive their share of the total charge.

It is admittedly desirable that the rates for line and radio transmissions should be co-ordinated. In a thoroughly integrated international telecommunication service the two systems would be interchangeable, for it is only by such interchangeability that full advantage can be taken of both. Although the capital cost of radio installations is much less than that of line systems, the development of radio is subject to limitations imposed by the number of frequencies available and there are also the interruptions due to atmospheric interference. Furthermore, new developments in line systems, particularly the greater use of coaxial cables and of submarine repeaters, point to an enormous extension of line channels available in the world. Only where line and radio systems work together as essential parts in an efficient telecommunication service can traffic, where necessary, be diverted from one to another to relieve overloading, and alternative routes be made available during periods of interruption.

Any proposed solution of high rates which looks to the exclusive use of less costly radio transmission for press messages is unrealistic not only in terms of the present composition of the world telecommunications network, but in the long view. But the co-ordination of the two kinds of transmission is clearly not compatible with the charging of different rates for the two services, since efficiency may often require full interchangeability between them.

Up to the moment, however, there are good grounds for asking whether this marriage of cables and radio has not led to the fixing of rates based on the most costly, and sometimes the least efficient, system of transmission. The most costly system for long-distance international news messages today tends to be transmission by cable. This may not always remain so. Current advances in cable techniques may ultimately make it possible to operate cable services more cheaply than radio services. What is open to question is how far such developments would, within the present rate structure, be allowed to work for the cheapening of international communications. For experience of the past makes it debatable whether the development of radio services much less costly than the older cable facilities has always been fully reflected in a lowering of the average cost of telecommunications as a whole. The experience of Press Wireless Ltd. suggests a considerable resistance to such adjustments by some of those whose approach to the problem has been shaped by the cost factors operating in older cable systems.

Press Wireless Ltd. came into existence following a decision by a number of American newspapers to meet problems arising from what they regarded as the bad service and high rates of many of the commercial carriers, by creating their own communication agencies. Because of the need to conserve short-wave frequencies, the Federal Radio Commission, forerunner of the Federal Communications Commission, insisted, however, on the formation of a single company.

"The effect of the new carrier on press rates," state Llewellyn White and Robert D. Leigh in *Peoples Speaking to Peoples*, "was almost immediate. Within a short time, the rates were cut from 50 to 80 per cent. Service was improved by the competitor and the competition alike."

The high rates now operating for press messages from many countries where fuller reporting of events and trends of opinion is desirable for international understanding, and the startling anomalies existing in the international rate structure, are serious obstacles to the satisfaction of those public interests to which the press and the telecommunications services should alike be dedicated.

Before considering how far this situation can be remedied, let us see to what extent similar obstacles affect other major factors in a sound press telecommunication system. For it is clear that no single part of the world telecommunication structure can be considered in isolation, and that no reform can be effectively implemented without full cognizance of the requirements and potentialities of each.

IX. *leased wire services*

The high cost of cabling news by ordinary commercial channels can and does affect the whole balance of world reporting.

As the already quoted comments from editors in various countries demonstrate, such costs help to make it impossible for many newspapers to maintain their own correspondents in important news centres, or to send visiting correspondents to cover significant current developments to the extent their own wishes and the public interest would otherwise dictate.

Nor can the services of even the major news agencies, however wealthy and strongly supported by newspaper members and subscribers, entirely escape the effect of such charges. All news agencies have to work within a budget. Because of the nature of the demands which their world news responsibilities place upon them, it is almost impossible to prevent these budgets from coming under severe strain from time to time. A sudden, large-scale news development in some part of the world may throw the most careful budgeting out of balance and produce deficits which force a policy of retrenchment to restore the financial balance. In such circumstances, the most immediate and effective remedy is to cut down news reports from those areas where cable costs are most expensive; and, in particular, to apply such economies to secondary but vitally important background dispatches in favour of flashes covering only hard news. When rates are extravagantly high, as many demonstrably are on the evidence already quoted, the temptation to do this is almost irresistible.

The effect of high rates upon the individual correspondent must also be taken into account. He knows that he must not without good cause exceed the budget allotted to him. But he also knows that he must cover hard news of importance whatever the cost. Where cable rates are high, he is therefore under constant temptation to let pass material which would give depth and background to the news, simply because it would be too expensive to use. In order to help international understanding it may be desirable to quote local press comment, to report views of representative men and women in interviews, to explain the social, political and economic background against which events ought properly to be seen. The good correspondent recognises this need. But if every word he sends is going to cost 1s. or 15 cents, he dare not risk dipping far into his budget for it. He has to bear in mind that although the long-term effect of such background news may be extremely important, it is unlikely to be immediately missed if he does not send it, whereas a failure to report hard news will be.

Thus the whole pressure of high cable rates tends to distort the picture of the world presented to the ordinary newspaper reader. First, by making it impossible for many newspapers to maintain correspondents abroad to the degree necessary to give a full individual service to their readers. Second, by placing upon even the most powerful agencies and their correspondents a well-nigh irresistible pressure to concentrate on the most dramatic and sensational world news instead of reporting events and trends which, although secondary in immediate news value, may be essential to a proper understanding of affairs.

Thus a lack of balance in the general structure of cable rates is almost inevitably reflected in the pattern of world reporting.

A comprehensive world system of news exchange depends, however, not only upon access to the maximum amount of information, but also upon an absence of all avoidable obstacles to the redistribution of information across frontiers by the most rapid, effective and economical means.

In highly-developed and highly-populated land areas with a well-established press, experience has shown that the most suitable means for this purpose is the international teleprinter network, composed either of line systems or of a combination of lines and point-to-point radio links.

It is the most efficient because it establishes direct contact between the distributing agency and receiver, and facilitates a two-way traffic, so that news can be sent and received and a true exchange of information made possible.

A news service by multiple address broadcast on a wide beam must, by the nature of the transmission employed, be the same for all recipients. A service sent by direct teleprinter links can, on the other hand, be tailored to meet the expressed needs of an individual newspaper. The subscriber receiving the service can "talk back" to the news agency. In this way, an individual editor may ask for further elucidation of a specific point in the news. He may feel that from the viewpoint of his readers' interests, he needs more information on a particular event than the agency's general service contains—for example, a full report of a speech by a statesman instead of a summary.

He can ask for this over the same lines by which he is receiving his service, and in a matter of minutes will have what he needs.

Thus, the international teleprinter network, directly linking the head office or regional distributing point of a major agency with a large number of individual subscribing newspapers and national agencies, provides one of the most effective practical replies to the danger of mass-produced news reports which many feel to be inherent in the concentration of world news services in a comparatively few hands.

It provides both the national news agencies and individual editors with a more effective voice in the balance of world news presentation than they would otherwise have, and at the same time greatly increases the ability of the major news agencies to draw upon many different national sources in the composition of their news services.

It was these advantages which led to the rapid post-war development of agency networks in Europe, a development pioneered by the American agencies, with their long experience in the use of transcontinental networks, but now followed by all. The expense involved in leasing circuits for a comprehensive European network is considerable. It may well increase an agency's costs by as much as $140,000 a year over the sum involved in serving the same area by multi-address broadcasts. But agencies and their subscribers have found that the superiority of the service thus made possible more than justifies the additional expense.

Leased teleprinter networks developed much later in Europe than in the United States, Canada and even India, which, despite its less advanced press system, had by 1940 a teleprinter network operated by Reuters, later taken over by the Press Trust of India and now extending over 24,000 miles. Delayed development in Europe was not due to lack of physical facilities, but to problems presented by national frontiers, the gearing-in of different systems and agreements over tariffs.

Some of these difficulties still persist and stand in the way of an even greater use of leased teleprinter networks on the European continent, and unless they are solved, they may also impede its later extension to other areas.

Apart from Spain, Portugal, Yugoslavia and Greece, where shortage of lines and the cost involved in the long haul necessary to link them with the rest of the European network affect the position, Western Europe does not lack the physical facilities for a still greater extension of international teleprinter networks and for the use of leased lines for other press purposes. But such an extension, which would offer so many advantages in the international news field, is obstructed by a number of anomalies which call for careful review by telecommunication administrations.

One of these lies in the fact that the charges made by different telecommunication administrations for leased lines exhibit divergencies comparable to, although not as great, as those in press cable rates.

The extent of these divergencies was considered by the European Technical Conference of Press Agencies at its 1952 meeting, which was attended by agencies of 15 European countries: AFP, France; ANI, Portugal; ANP, Netherlands; ANSA, Italy; APA, Austria; ATS, Switzerland; Belga, Belgium; DPA, Federal Republic of Germany; INA, Eire; NTB, Norway; Reuters, United Kingdom; Ritzaus, Denmark; FNB, Finland; Tanjug, Yugoslavia and TT, Sweden.

The conference had before it a paper by Cecil Fleetwood-May, European general manager of Reuters, urging joint action to bring before the ITU's International Telegraph Consultative Committee the danger of telegraph administrations using their monopoly position to charge high rates for the leasing of lines to agencies. Practically all the agencies represented agreed that although differences in the rates for leased lines charged by various administrations might be partly due to the nature of available lines and the terrain over which they run, and might also be affected by differences in living and other costs affecting all prices, they nevertheless appeared in large measure to be purely arbitrary. The conference therefore appointed a committee consisting of Cecil Fleetwood-May, Reuters; Dr. Siegfried Frey, general manager of Agence Télégraphique

Suisse (ATS); Maurice Nègre, director-general of Agence France-Presse; Count Riccardi, vice-president of the Agenzia Nazionale Stampa Associata (ANSA), and Colonel Olof Sundell, general manager of Tidningarnas Telegrambyra (TT) to examine the question further.

The extent of the discrepancies in rates may be judged from the following list which was submitted to the conference. It shows the variation in rental in nine European countries for a typical 300 kilometre duplex teleprinter line at the lowest rate available to the press. The rate in Norway, the country with the cheapest tariff, is used as index factor 1. Norway, 1; Denmark, 1.05; France, 1.23; Sweden, 1.25; Holland, 1.50; United Kingdom, 1.52; Germany, 1.79; Belgium, 2.82; Switzerland, 3.43.

The high Swiss rate, it may be noted, is rivalled outside Europe by that in Egypt where the index figure is 3.37. The rate for India works out at only 1.35. It is difficult to give a comparable figure for the United States because of the fact that under the American system the cost of leasing a line is based on a sliding scale according to the time of day. But the rate even for the peak period would seem to be below the majority of European rates.

Although it is difficult to find a good reason for such wide divergencies from country to country in the cost of leasing lines, there is an even more serious anomaly where international teleprinter services are concerned.

This arises from the discrepancy between national and international rates for leased teleprinter lines in Europe. Such lines have become the standard channels for the exchange of news in Western Europe. Yet owing in part to the linking of international telegraph rates with a purely nominal gold franc, it can and does happen that a teleprinter line, available at a certain rate per kilometre between two points within one country, is charged for at a much higher rate for identically the same line if it is electrically connected to a wire going over the frontier and thus becomes "international".

An agency teleprinter wire between Paris and Marseilles is, for example, charged for at the internal rate. But if it is desired to carry on this same line by an extension to Madrid, then the line at once becomes "international" and will, under existing international telegraphic agreements, be charged for as such all the way from Paris. Conversely, an agency with an international line from Paris to Madrid, passing through Marseilles, would not be allowed, under existing telegraphic regulations, to drop off copies of messages to a subscribing newspaper in Marseilles, because that would be using an international line for internal communications.

This situation is made still more complicated by the fact that the value of the gold franc in terms of present national currencies is subject to arbitrary decision and may vary even within one country and one telegraphic administration. In Sweden, the telegraph administration has a high international reputation for its sympathetic approach to press problems and charges. It charges one of the lowest rates in Europe for internal teleprinter lines. Despite this, the national agency, Tidningarnas Telegrambyra, which has teleprinter links with agencies in the other Scandinavian countries, finds that rates for international lines are sometimes calculated on the basis of the gold franc being equal to 1.50 Swedish crowns and at other times give it a value of 1.69227.

Similar or greater anomalies exist in pratically every European telegraph administration.

The best way to avoid the anomalies arising when the same stretch of line is charged for at a national rate in one instance and at an international rate in another, would seem to be for each administration to fix an economic rate for the use of lines within its own frontiers, without any question of an over-all international rate. Such a system of separate charges by each administration concerned in international transmission is not practicable for ordinary telephone or telegraph toll traffic. But there would seem no reason why it should not be adopted in the use of leased channels, where quite other conditions exist. It would mean a big reduction in present international charges, while it would leave each administration with an economic return on the lines used, since it would receive the same return for this international employment as that regarded as satisfactory when they were leased internally.

This system is in fact used by the British Post Office. For a line to the Continent, the Post Office charges its own domestic tariff up to mid-Channel, leaving the administration at the other end to charge its own rate for its own section of the line. There would seem no good reason why this practice should not be generally adopted.

Although the world news agencies are the main press users of leased lines and radio channels, they are not the only ones. The renting of such channels for the coverage of particular events is a necessary part of radio news services too and here again all kinds of anomalies can be found.

A radio news service is expected not only to put out news bulletins received from the agencies or cabled by correspondents, but also to broadcast live reports from its men covering important news. The cost of such reports is often, however, artificially inflated by the imposition of an excessive minimum period for the hiring of channels, plus a compulsory test period which must be paid for.

The minimum period bookable for radio news transmission from Delhi, for example, is 15 minutes, plus a test period of 20 minutes, both of which must be paid for. As a result, the cost of a single short news dispatch in voice from Delhi to Europe is brought up to $224, which makes the use of radio for direct news reporting from India quite impracticable except in very special circumstances.

For European news broadcasts a 15-minute test period is normally imposed, which must be paid for in addition to the transmission period. But in the United States, a radio channel can be booked for as little as 10 minutes, with no compulsory test period. Where, however, a news broadcast from there involves the use of internal telephone lines to carry the programme to the transmitter, a minimum booking of one hour is imposed and the cost rises with the distance. For a news broadcast to Europe from Washington, it is $34 an hour; for one from Chicago it is $150, and from San Francisco $400.

Since radio transmissions involving the use of such lines are rarely likely to last more than 15 minutes, there is in every case an artificially imposed wastage of time and money.

This system of imposing minimum transmission time, plus in many cases an additional test period in excess of that actually required, not only increases the cost of radio reporting, but often imposes an unnecessary and quite artificial strain on telecommunication facilities. This strain can be sufficient to cause serious delays in the transmission of normal press messages.

During the 1952 Olympic Games at Helsinki and the NATO Conference in Lisbon, for example, these requirements greatly increased the competition for communication channels. At the 1946 Moscow Conference, they caused a serious holdup in press copy which had to wait because radio services were forced to monopolize channels booked in advance for much longer than they were actually needed.

Apart from such disparities and anomalies in the structure of rates, there remains one important question of principle.

Under the present telephone and telegraph regulations, if a line is leased for use by more than one user, it becomes a "joint line" and as such is charged for at a substantially higher rate.

The purpose of this provision is to prevent those leasing private lines from making a profit out of them by sub-letting to others in competition with the normal business of the telegraphic offices. A brokerage house enjoying the advantage of a private line between two important market centres has been known, for example, to enter into the telegraph business itself and run a service for other brokerage firms.

Regulations to prevent such practices are fully justified. But as at present applied in the press communications fields by many administrations, they work directly against the international public interest which requires the maximum development of exchange agreements between agencies.

Under existing regulations, a world agency is perfectly free to lease a teleprinter line to serve its fee-paying clients at the normal rate. But the position is immediately changed if, as part of that movement towards the development of international or regional news co-operatives to which reference was made in an earlier chapter, the world agency should wish to come to an exchange agreement with a number of national agencies to include the joint hiring of a teleprinter network for their mutual advantage. The position likewise changes if, in an area where common political, economic and geographical interests make the maximum exchange of information across frontiers desirable, a group of adjoining national agencies seek a similar pooling arrangement. In each of these cases the lines used for an admirable purpose would immediately become subject to a much higher rate, which might well place the service beyond the bounds of economic feasibility. Yet there could be no question of the telegraph administrations themselves being involved in any greater cost than if the same lines were leased by one user.

Nor is this the only instance in which a regulation designed for the legitimate protection of the telegraph services in certain circumstances can, when rigidly applied to press traffic in quite other circumstances, provide a serious obstacle to a full flow of important international news.

It is, for example, very desirable that adequate reporting of international conferences should be available to the press of all countries. Yet when two agencies recently arranged to share a leased line in order to provide a fuller coverage of the Lisbon NATO Conference than they could otherwise have afforded, they found themselves suddenly faced with charges which made the contemplated level of reporting quite uneconomic. The reason was a ruling that such a service must be regarded as bringing the line within the category of "joint use".

The application of this principle does not only stand in the way of an attempt by news agencies to undertake exactly that type of public service reporting which ought most to be encouraged. It presents a serious obstacle to a development which could be of great advantage to individual newspapers acting in association.

The high cable cost barrier which operates against the use of individual newspaper correspondents could often be substantially reduced by a co-operative agreement among the newspapers of one country, acting perhaps through their professional or trade association, to lease a circuit for a certain number of hours per day for their joint use. Such a method could be applied particularly to newspaper coverage of important United Nations meetings or other international conferences. But it could only be applied if it were possible to lease a line at an economic rate, unaffected by its present restrictions to one user.

Where an international circuit involves the use of a radio link subject to fading, such a system would have a further advantage. It is no part of the business of an ordinary telegraph administration to discriminate between the relative value and urgency of particular press messages, nor is it competent to do so. It can merely transmit them in the order in which they are filed.

Yet, as in the case of the Australian radio beam from London, the fading of a radio link, often for long periods, and the lack of any order of priority for press messages may almost completely offset the advantages of even the cheapest cable rates.

Such difficulties could be overcome in certain circumstances if national press associations could arrange the co-operative leasing of a radio or line circuit and establish their own joint editorial unit to handle the order of transmission according to agreed news priorities. This was done on a large scale at the time of the "D-Day" landings in Europe, when the British Ministry of Information and SHAEF Headquarters jointly set up an editorial telecommunications unit of this kind in London to handle the transmission of all war correspondents' dispatches. Without this unit the news could never have been filed at the speed it was. Such a system, on a smaller scale, is perfectly applicable to many peace-time news enterprises. Its adoption was, in fact, discussed at the last British Commonwealth Press Conference, when the possibility of an increase in the Commonwealth press rate was foreseen. The Commonwealth Press Union has set up a committee to consider the practicability of such a scheme.

This arrangement might be more widely considered by national press organizations whose members cannot now provide their readers with that service of foreign news and comment by their own correspondents which is needed to supplement the basic news service of the world news agencies.

But no such development is possible without a substantial amendment of present telecommunication regulations regarding the use of leased lines.

x. multiple address newscasts

International teleprinter networks provide the most complete and satisfactory answer to problems of the exchange of news across frontiers in highly-populated and economically-developed areas. There are many parts of the world, however, where the technical facilities for them do not now exist or are not likely to do so in the foreseeable future.

But this does not rule out the possible extension of international teleprinter networks.

A study is now being undertaken for the ITU by the Joint Committee for the General Switching Programme of the International Telephone Consultative Committee (CCIF) to determine the best way of connecting Middle Eastern and South Asian countries with the network of major international telecommunications lines in Europe and the Mediterranean basin. This effort may, if given adequate financial and technical support, eventually open up great new areas to such developments.

Present technical advances in the use of coaxial cables and in transmission methods may, moreover, produce an immense change in the old telecommunications pattern by vastly increasing the number of channels available for all kinds of traffic.

Whereas the number of channels available on the older telegraph and telephone lines was severely restricted, a modern two-pair coaxial cable, though smaller in size, can provide 960 telephone channels and 24 times that number of telegraph channels. Indeed the number of rapid, efficient and relatively cheap channels which could be made available for international communication would, technically speaking, appear to be almost unlimited.

This technical advance, taken in conjunction with the development of submarine repeaters capable of greatly increasing the load carried by ocean cables, points to a possible material expansion in world telecommunication facilities. This may rival and even exceed the immense expansion which followed the invention, first, of the electro-magnetic cable and secondly, of radio—developments which have changed the face of the world in the past century.

This forecast is not based on any expectation that new and startling techniques will be discovered, but upon an evolutionary progress in design and technique which one telecommunication authority has likened to current developments in aerodynamics. In his opinion, the introduction of voice-frequency working, coaxial cables and regenerative repeaters represents an advance comparable to that heralded by the jet engine. But the practical impact of this advance and of technical developments in facsimile transmission is yet to come.

Many recent improvements in technical practice are only just beginning to make themselves felt—particularly in the field of international long-distance traffic—with the re-laying, repair and modernization of war-damaged cables and the laying of new ones. The increase in capacity made possible by modern techniques has not yet, therefore, been fully realized.

The repair of existing cables and the laying of new ones is expensive, and therefore bound to take longer in a period of rising costs. Nevertheless, it is likely that, before long, total cable capacity will draw level with and surpass the amount of traffic now offered. Eventually, increased capacity may make possible an enormous expansion in world press traffic at economic rates to meet the ever-mounting demand for information.

This development in line communications is being accompanied and even overtaken by comparable advances in radio techniques, particularly in the development of "printerized" radio. By this is meant radio-teleprinter facilities which equal the wire teleprinter in speed and efficiency for one-way traffic, but are less costly over great distances.

Many of the new telecommunications techniques now being developed can be applied to both cable and radio. It seems certain that the world telecommunication pattern will increasingly be one of co-ordinated radio and line systems in which channels will sometimes be wire and sometimes radio. On the other hand, multiple address transmissions by wide beams giving umbrella coverage over vast areas are likely to be for some time, and perhaps in some instances permanently, the most suitable and economically viable means of distributing news to, or receiving it from many comparatively undeveloped countries. A steady flow of information between such areas and the main news centres of the world is of vital international interest. It is therefore important that adequate facilities for transmitting and receiving such newscasts should exist, and that their operation should not be handicapped by unnecessary restrictions.

These facilities are however, still limited; a recent ITU report showed that there are only 14 countries offering facilities for radio communications to several destinations. Moreover, the extent of these services and the method and level of charging for them varies considerably.

The extensive facilities made available by the British Post Office, which pioneered in this field, and which leases transmitters to Reuters for seven Globe-reuter beams capable of covering almost the entire world, have already been described. In the United States, Mackay Radio, Press Wireless, Radio Corporation of America and the Tropical Radio Telegraph Company provide multiple address facilities, charges being computed on the basis of transmission time. The big American news agencies use these services extensively, although not so extensively as Reuters, and the accusation made against them in *Peoples*

Speaking to Peoples in 1946 that they were failing to shoulder their news responsibilities in this respect is no longer true to any great extent.

The authors of this report, Llewellyn White and Robert D. Leigh, then declared their belief that: "The acid test of good faith in so far as the press associations are concerned is their willingness to exploit multiple address newscasting to the extent that Reuters, no less jealous of government interference, has exploited it for four years.... [We] find it difficult to reconcile the AP's and UP's promises to the American people with the fact that, three months after its offer of worldwide multiple address facilities at one-third of a cent a word, the Mackay Radio Company had no takers."

Although this complaint no longer holds good of the big American agencies, it is quoted because it well expresses the sense of public responsibility which should motivate the business of news distribution. This responsibility can be met only if the principal news agencies are ready to undertake news distribution to every quarter of the globe, irrespective of the relative profitability of one area compared with another. Within the terms of their public responsibility, the agencies cannot be selective and exclusive.

What is true of the news agencies is equally true of telecommunication services, and this should be borne in mind in considering the facilities they are prepared to offer.

Apart from the United Kingdom and the United States, the largest operator of multiple address newscasts is France. Senders and addressees must be authorized by the French Ministry of Posts, Telegraphs and Telephones, which charges for short transmissions on the basis of a rate per word varying with the power of the transmitter and the number of addressees, and for long messages on the basis of transmission time.

The Netherlands also has adequate facilities for this type of news service. Transmissions are carried out exclusively by the Ministry of Posts, Telegraphs and Telephones which bases its charges on the time and speed (words per minute) of the transmission. The Federal German Republic has also developed multiple address broadcasts extensively.

Only four other European countries offer these facilities. They are Poland, where the charge, at so much a word, is increased if a message is sent to more then five addresses, Portugal, where charges are based on monthly wordage totals, and Belgium and Switzerland, which both calculate transmission fees individually, according to the circumstances of each case. In Italy, no service is provided by the administration, but a news agency has been authorized to operate one. The administration will receive a percentage of gross takings.

Austria, Denmark, Greece and Sweden do not offer transmission facilities but have organized reception services. Spain has a service under consideration. The other European countries replying to the ITU request for information were Albania, Czechoslovakia, Finland, Hungary, Iceland, Ireland, Luxembourg, Norway and Turkey. They provide no facilities either for transmission or reception.

Apart from Europe and the United States, 35 administrations sent information to the ITU. Only seven of them have authorized transmission and reception services.

These are Argentina, where transmission charges are fixed on a time basis; China; Malaya, which makes a rental charge for equipment and hire of transmitter time; Egypt; Indonesia, where the charge is related both to the duration and speed of transmissions; Paraguay, and Viet-Nam.

Australia has no transmission services at present in operation but is prepared to admit them. Although there is no transmission service in Pakistan, the AP and UP have been authorized to receive multiple address broadcasts for distribution to their subscribers.

It will thus be seen that the multiple address system has not yet won universal acceptance, despite its great international value. Even where it exists the cost may vary considerably. Authority to operate it may sometimes be allowed to news agencies under licence. In other cases it is restricted to the adminis tration itself.

The same is true of reception. In some countries, authorized news agencies or newspapers are permitted to receive multiple address newscasts directly. In others, the authority to receive is reserved for the administration. The charge for reception varies greatly. In Austria, it is five schillings per 100 words, mounting to 150 schillings for over 5,000 words, with a double charge for dispatches originating outside Europe. In France the charge for the receipt of multiple address telegrams is the same as for ordinary radio-telegrams. In the Netherlands, addressees, who must be authorized, are charged a flat fee of 500 florins per annum. In Greece a flat rate of 500,000 drachmas is levied on news agency or newspaper recipients; in Sweden there is an annual receiving licence fee of 5,000 crowns; and in Portugal a monthly subscription charge, according to wordage received.

Outside Europe, there are even greater variations in practice and cost. In many areas, moreover, no reception or distribution facilities are provided— sometimes, it would appear, as a matter of deliberate policy; in others, restrictions are imposed which gravely limit the value of the services.

In this connexion, a statement by Cecil Fleetwood-May, Reuters' European manager and a pioneer in the multiple address field, is significant. Describing Reuters' experience when it first sought to develop its Globereuter service of worldwide newscasts, he said: "A great deal of my personal effort on behalf of Reuters over the application of multiple address radiotelegraphy to news distribution has been used up in fighting in one country after another the imposition of artificial and restrictive receiving conditions. The conditions which telegraph authorities sought to impose were designed to maintain the charges and limitations associated with the older and slower methods of sending press material over point-to-point radio and cable circuits."

The principal world news agencies have, in general, succeeded, though only with considerable difficulty and after much labour, in removing many of the obstacles to multiple address reception which originally hampered its development. But some of them still persist.

The artificiality of much of this opposition and its lack of relation to any sound budget principle is illustrated by the following fact. In some areas, where the multiple address system is now accepted, the existence of only one receiver of a news agency service has recently more than once been used to substantiate the claim that multiple address conditions are not being met. The news agency, it is claimed, must therefore pay more for distributing its service to one subscriber than to a number.

Further expansion of the multiple address system depends upon much broader provision of adequate and unrestricted reception facilities.

These facilities could be provided by the world news agencies, by national agencies or by individual newspaper subscribers. In some cases they already are. But in many other instances, reception arrangements are still forbidden by the telecommunication administrations, or only allowed if subject to charges which seriously limit the availability of the service to those who may need it most.

Reform in this field would seem to be a matter for serious consideration by all members of the ITU.

The United States delegation to the International Telephone and Telegraph Conference of 1949 pointed the way to such reform when it asked that a declaration be placed on record that the United States would not use the authority of the International Telegraph and Telephone Regulations to restrict the reception of press radio communications. The delegation also expressed the

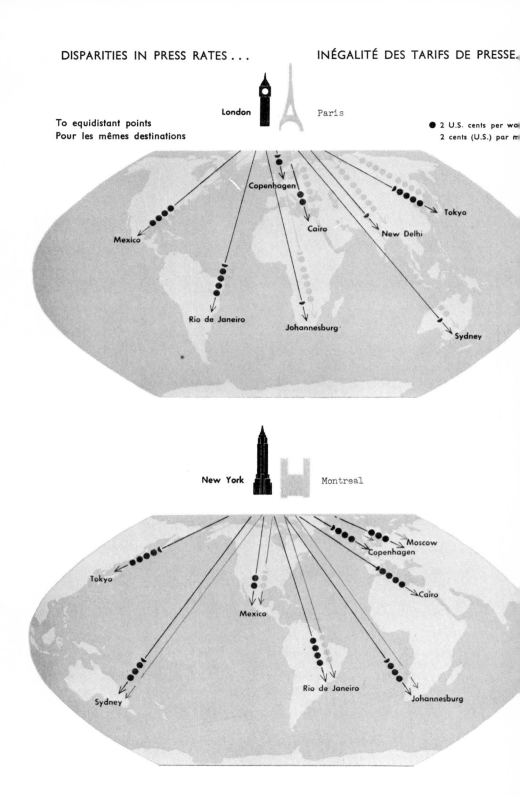

DISPARITIES IN PRESS RATES ...   INÉGALITÉ DES TARIFS DE PRESSE.

London   Paris

To equidistant points
Pour les mêmes destinations

● 2 U.S. cents per wo
  2 cents (U.S.) par m

Copenhagen

Mexico

Cairo   New Delhi   Tokyo

Rio de Janeiro   Johannesburg   Sydney

New York   Montreal

Tokyo

Moscow
Copenhagen

Mexico

Cairo

Rio de Janeiro

Sydney   Johannesburg

On identical two-way routes
Dans les deux directions

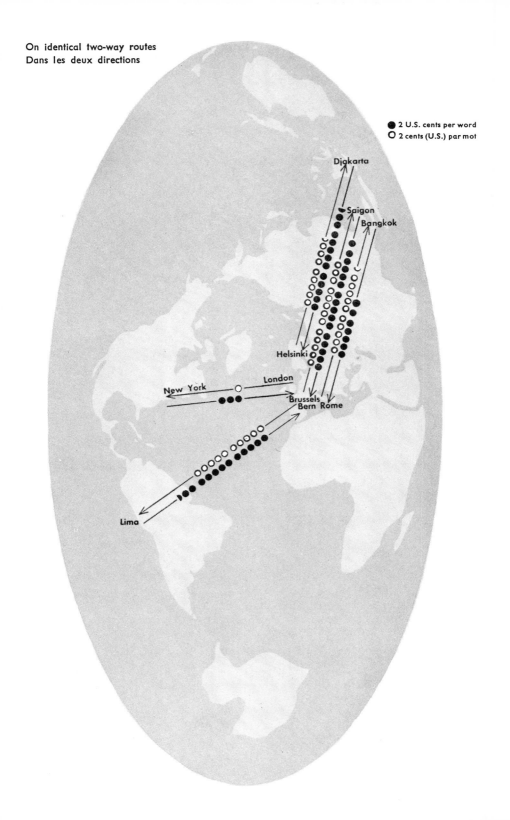

● 2 U.S. cents per word
○ 2 cents (U.S.) par mot

Djakarta

Saigon
Bangkok

Helsinki

New York — London

Brussels
Bern Rome

Lima

hope, which was published in an appendix to the regulations, that other administrations would follow a similar course.

This hope has not, however, been generally fulfilled. Until it is, serious obstacles will continue to hinder the full use of a news distribution system which has immense potentialities for public information and international understanding.

XI. *facsimile and telephoto services*

In the last resort, all systems of international news distribution, whether by point-to-point line and radio transmissions, by international teleprinter network or by multiple address newscasts, depend for complete efficiency on the communications systems within the area of receipt.

In countries where adequate internal distribution facilities are lacking, international news from the world's main centres can reach only a comparatively few towns and will never get to the majority of populations.

Highly-developed newspaper areas, on the other hand, and particularly most European and North America countries, have seen such distribution facilities reach a high pitch of efficiency. Internal teleprinter networks and, in some cases, telephoto networks link not merely all the principal cities but every town where a newspaper is published or a radio station operates.

In such circumstances, news received at a big distribution centre can be disseminated almost simultaneously to every part of the country. Each section of the population can thus be offered an equal service of national and world news.

Even in areas where newspaper development has not reached the same mechanical and professional level, similar if less extensive telecommunication services make rapid news distribution possible. In Egypt and India, for example, the main agency news services can be distributed to subscribers by teleprinter over a fairly wide radius.

Such conditions, however, do not prevail in many areas. One of the biggest tasks of telecommunications is to remedy the immense disparities in physical facilities which obstruct comprehensive exchange of world news.

In this task radio has an important role. In many areas where the cost of developing line systems is prohibitive, radio networks can be developed within limits of expenditure which the standards of public interest would amply justify. The United Nations Technical Assistance Programme, in which the ITU is now participating, could be of enormous help here, provided national administrations sufficiently appreciated the importance of the issues involved.

Great technical advances in the whole telecommunication field are still being made and may continue to be made at increasing speed. Meanwhile current technical achievement has made possible a revolutionary change in the means to increase international understanding in many areas if the will to use those means exists.

In many instances, the chief obstacles lie not in any lack of physical and technical facilities, but in mental attitudes which frequently derive from out-dated methods or practices and raise a solid wall of resistance to change and development, even when there is abundant evidence of the advantages obtainable from them.

The high press telegraph rates charged in many Middle Eastern and other countries, for example, often appear to be due as much to obsolete ideas in administration as to obsolete or scarce equipment. The high rates do not bring high returns, because they keep the volume of traffic too low. As a result, it is frequently argued that they cannot be reduced because, high as they are, they yield such a small return. Yet the whole body of experience in other countries shows that moderate telecommunication rates almost invariably stimulate an increase in traffic that sharply increases receipts. As regards press traffic specifically, the lesson is that where rates are high, service is poor, but where rates are moderate, traffic is abundant and service usually good.

Ultra-conservative mental attitudes tend not only to keep telecommunication rates unnecessarily high, but to obstruct new developments. There have been instances in the past where not merely the financial, but what may be described as the mental investment in cable systems has hindered the full use of oppor-tunities provided by technical advances in radio communication. This no longer appears to be true of major telecommunication operators, whether public or private. Radio and line systems are generally being co-ordinated and each used to the best advantage as part of an integrated service. In the broadcasting field, however, the full development of frequency modulation radio (FM) has almost certainly been delayed through opposition from groups interested in the older amplitude modulation (AM) system.

The development of facsimile transmission may be in some danger of running into similar obstruction. Resistance here is not confined to telecommunication services committed to older techniques, but is also found among some news-paper groups.

A facsimile transmission system by which text or photographs or both could be sent by radio was developed by Austin C. Cooley as long ago as 1926. A few years later John V. L. Hogan and W. S. H. Finch succeeded in carrying out facsimile broadcasts of a considerable technical standard. By 1937, six stations in the United States had been granted experimental licences to broadcast printed news and other material during the night hours. In 1948 the Federal Com-munications Commission authorized regular transmissions and the full com-mercial use of facsimile.

It is now perfectly feasible for a facsimile newspaper to be broadcast from FM transmitters to individual receivers. This can be done by means of facsimile recorders attached to or built into an ordinary FM radio set in the same manner as a record player. To receive such a newspaper, all that the "reader" needs to do is to tune in his receiver to the station radiating the facsimile signal. A copy of the facsimile transmitted (a page of news illustrated by photographs, for example) is then immediately reproduced on a chemically-treated paper unrolling under a printer blade on the home recorder; an up-to-the-minute radioed newspaper delivered straight from the press to the living room!

The Hogan Facsimile System, which could go into immediate commercial production if the demand were adequate, can achieve speeds of up to 500 words a minute. A refinement of facsimile experimentally developed by the Radio Corporation of America, and known as Ultrafax, uses a television transmitter instead of the FM transmitter employed by the Hogan system. Ultrafax has already achieved speeds greatly surpassing 500 words a minute and far beyond anything previously recorded. During a laboratory demonstration at the Library of Congress in Washington in 1948, the 1,047 pages of the novel *Gone with the Wind* were transmitted by Ultrafax in 2 minutes 21 seconds.

Facsimile techniques are likely to be adapted to long-distance line transmissions in the future and offer great possibilities of more rapid communication. Some internal cables in the Federal Republic of Germany already use them but no international cable or telegraph systems yet do so. Facsimile's main commercial development in its country of origin, however, has so far been for short-distance communication in inter-office systems and as a pick-up and delivery service for telegrams by Western Union. Since the war, more than 2,000 "desk-fax" units have been installed by Western Union in customers' offices for the transmission of telegrams to the nearest Western Union office. These units are fully automatic. The customer merely places his message in the machine and presses the starting button.

The most revolutionary potential field of facsimile development lies in the broadcasting of newspapers directly from press to home. Limits are here imposed, however, by the fact that broadcast facsimile requires the use of FM or television transmitters using very high frequencies which are restricted in range.

This would not matter greatly in areas with a sufficient number of stations to give a fairly wide coverage—notably the United States, where most of the original development work has been carried out. But even there, facsimile has so far failed to win adequate commercial backing to make large-scale broadcasting possible. The advantage of immediacy offered by the facsimile newspaper may, indeed, prove insufficiently attractive to the general public in highly developed newspaper publishing areas to bring about the change in reading habits which large-scale commercial operation would require. Various newspapers, including the *New York Times*, the *Philadelphia Bulletin* and *Philadelphia Inquirer*, the *Baltimore News-Post* and the *Miami Herald* have, nevertheless, carried out facsimile demonstrations.

Failing the existence of a chain of FM or television stations capable of retransmitting printed pages from one to the other, the international facsimile newspaper envisaged by some of those concerned in early developments does not seem a practical possibility. It might well become so, however, if the demand for it were to grow sufficiently strong.

The possibilities of this system in underdeveloped newspaper areas, where the use of radio facsimile might facilitate a wide distribution otherwise unattainable without heavy investment in printing plant would, however, seem worthy of close examination, provided the obstacle of high cost of receivers could be overcome. This obstacle might be surmounted by establishing co-operatively-owned group receivers in areas without their own newspapers. On the transmitting side, FM equipment is much less expensive than that used for regular AM broadcasting, and because of the limited range of the ultra-short waves employed, the risks of mutual interference are far less than with AM transmission.

The facsimile newspaper is a matter for the future but it is essential to ensure that no avoidable obstacles should hinder its development, should it prove feasible.

Almost as urgent as the removal of impediments to the maximum transmission of world news is the need to extend facilities for the worldwide transmission of pictures by radio and telegraph—a need which facsimile might well help to meet.

Pictures provide an international language. The photographic divisions of the British Ministry of Information and the U.S. Office of War Information found that visual presentation can promote international interest among illiterate and semi-literate peoples whom the written word cannot easily reach. And as modern journalistic practice has increasingly demonstrated, it provides a valuable supplement to the written word for even the most sophisticated newspaper-reading public, bringing a sense of direct participation that the cabled report alone very rarely attains.

As with teleprinter facilities, the contrast between available telephotographic equipment in the most-advanced and least-advanced countries is enormous.

In addition, though adequate facilities for internal telephoto transmission have been developed in countries with highly-organized press and news agency services, international telephoto communication, even between important centres, has lagged behind.

This is partly due to causes outside the field of telecommunications. Although the European teleprinter networks used by the major agencies could be adapted to phototelegraphy and facsimile, pictures are generally exchanged in Europe over public telephone channels at a charge of so much per minute of transmission time. This is because newsprint shortages and the consequent small size of newspapers have so limited the demand for photographs as to make the leasing of permanent channels for picture transmission uneconomic.

As long as newsprint remains short, the use of photographs by newspapers in large areas of the world will continue to be restricted and the demand for permanent channels may remain small. This does not, however, affect the need for facilities to transmit photographs of particular significance or urgency.

At the present stage of press and telecommunication development it should be possible to take for granted the availability of such emergency facilities. To do so would be a great mistake. According to a 1952 report of the ITU, 13 European countries admit a photo-telegraphic service, and 23 other countries have organized such a service for extra-European transmissions.

Facilities in the two principal telecommunication centres, London and New York, far exceed those available elsewhere. In addition to the European links, telephoto services exist between London and 15 countries outside Europe. These include the British Commonwealth countries and the United States, Egypt, Israel and a number of South American States. In the United States, the Mackay Radio and Telegraph Company operates a photo-telegraphic service to Brazil, India and Israel. Press Wireless operates a similar service to France, Germany, Italy, Switzerland, United Kingdom and U.S.S.R. Radio Corporation of America operates an extensive service to 19 countries, including Australia, Austria, Brazil, Greece, Egypt, France, Germany, Italy, Japan, Korea, New Zealand, Philippines, Portugal, Sweden, Switzerland and the United Kingdom.

The French administration, in addition to its European links, operates telephoto services with Argentina, Egypt, Israel and the United States. The only telephoto channel into and out of Moscow is that operated by Press Wireless from New York.

Apart from facilities in the main telecommunication and news centres, most telephoto services are comparatively small and restricted in scope. An exception is facilities in Argentina, which has developed services to the United States, Europe and Brazil.

Of the 69 countries covered by the ITU report, more than 30 have no facilities of any kind for sending or receiving telephotos. The total of countries without such facilities in the world as a whole is considerably greater.

Barriers to the full exchange of news expressed in pictures are thus still very great, even greater than those to news in words. Despite the immense technical advances already achieved in telecommunications, and the even greater ones which can now be envisaged, we are yet far from that full and free flow of information between peoples which international understanding makes increasingly imperative.

part three

XII. conclusions and recommendations

It was suggested at the opening of this survey that the free flow of news across frontiers is a major international interest transcending the professional interest of those engaged in its collection and distribution. Without knowledge of peoples by peoples, peace and understanding are made more difficult in our complicated yet shrinking world. Without news of each other, we live as strangers among our fellows.

By its nature, a survey such as this must in many respects be tentative. The area of telecommunications is so wide, the number of administrations and agencies involved so numerous and their problems, while having a common basis, differ so greatly in detail, that any attempt to review them will doubtless seem partial and incomplete. In so complicated and far-flung an activity, it is difficult to acquire all the facts necessary for a comprehensive and objective study. Yet, even if those directly concerned in overcoming telecommunication problems should feel that the reasons for what appear to be anomalies, discrepancies and deficiencies have been incompletely presented, the broad picture that emerges can hardly be disputed.

This picture, remarkably favourable in many respects, reveals a high level of technical achievement and public responsibility among those now charged with leadership in the field of telecommunications. But it is a picture which, by reason of circumstance, historical development, and economic and other factors, inevitably has some grey and some black patches. It shows, for example, that administrations have not always kept pace with the possibilities inherent in technical achievement, or with the changing requirements of public interest. Such time-lags are not exceptional. They are to be found in all large-scale human enterprises. But their effects can be particularly important in the field of mass communication.

Conclusions drawn and recommendations made in this chapter are personal, although they are the fruit of detailed discussion with numerous operators and press users of telecommunications systems. Not everyone will accept as equally valid everything that is here suggested.

Yet I believe it is indisputable that, on the evidence available, a strong case exists for a thoroughgoing re-examination of some of the principles and many of the practices that now mark the field of press telecommunications. Such a re-examination can only be effective if it is founded on a firm appreciation by press services, telecommunications services and governments of the vital public interests involved. In many cases where reforms are most urgent, these interests must be given pre-eminence over strictly budgetary and commercial factors, however necessary it may be to bear in mind the limitations which such factors impose upon ideal solutions.

In the vast complex of international telecommunication services, those used for press purposes are only a small sector. Although the total volume of press messages transmitted around the world by cable and radio, telegram and telephone, teleprinter network and multiple address radio is quantitatively large, is still growing and should continue to increase, it represents only a relatively small part of the total business handled by the world's telecommunication services.

Yet, in terms of public interest, it is among the most vital of all the activities with which those engaged in telecommunications are concerned. This was recognized many years ago by the administrations represented in the first International Telegraph Union. The struggle for improved telecommunications during the past century or more has been as much a part of the struggle for press freedom as any of the battles against censorship.

In that struggle to widen the area of communication between peoples, the press and the telecommunication services have again and again been allies, although their alliance has not always been free from misunderstanding. At times, the pressure of newspaper demands and the exercise of newspaper initiatives have brought about advances in telecommunication methods and techniques, which, failing these incentives, might have been long delayed. At other times, new discoveries and far-ranging developments initiated by telecommunication administrations have opened the way to great advances in the exercise of press freedoms.

This alliance between telecommunication services and the press is founded not only on the mutual interest shared by suppliers and users of an important service, but upon something even more vital: their common participation in a public responsibility which is basic to our civilization.

Over the greater part of the world, the common pattern of press organization is that of private ownership, governed in many of its operations by considerations of profit and budgetary solvency which exist in all commercial enterprises. Similar considerations rule in a number of large and important privately-operated telecommunication services. Even where public ownership prevails, as is now the case in most of the field, it operates according to principles of profitability and the maintenance of budgetary balances which must be fully considered in managing services and fixing rates. Nevertheless, press and telecommunication services have public responsibilities greater than their purely commercial interests, and fully recognize the fact. Their acceptance of these responsibilities gives them the national and international status they enjoy. The obligations they thus assume call for judgments on what is desirable and possible, in which weight must be given to considerations going far beyond budgetary factors. The extension and improvement of press telecommunication services so as to facilitate a fuller and freer flow of news depends greatly on the acceptance by those concerned of such non-commercial standards.

In the field of international news exchange, the responsibility imposed upon newspapers by the whole concept of a free press requires them to develop their international correspondence to the maximum extent, even where the maintenance of foreign correspondents may involve expenditure which would not be commercially justifiable if immediate financial return were the sole consideration. The concept of a free press is based not on any belief that its organs, as such, should enjoy a special status, but upon the right of the individual citizen to receive all the information necessary for an independent judgment on affairs, and the consequential right which newspapers claim in order to satisfy this demand. The validity of these rights depends on the acceptance of obligations greater than those which stem from the commercial need for profitable circulation.

To fulfil their responsibilities within this concept, newspapers must serve the public interest through honest and comprehensive reporting, even when a greater financial return could be secured by economizing on world news services which may fail to bring immediate reward in higher circulations and increased profits. Newspapers must pay their way. But they must be prepared to include in their overall budgets a public service of world news reporting which may be relatively much less profitable than national news and feature services. They must accept the obligation to give within their available space a balanced and objective picture of world events, even when concentration on dramatic and sensational items may seem more attractive by standards of immediate circulation appeal.

Only if newspapers are prepared to undertake such obligations to the fullest possible extent do they have the right to demand special consideration for the transmission of news. Only if they show themselves ready to use all the available facilities for cheaper news transmission with the object of further developing "public interest" reporting can this demand be made with genuine assurance.

Upon the world news agencies this conception of public responsibility imposes a special duty—the obligation to report news from and to less-developed and relatively less-profitable areas of the world, as well as to those where large newspaper concentrations bring substantial returns. Performance of this duty would seem to require much wider extension of the principle that the charge for a comprehensive news service should be equated to the circulation of the newspaper receiving it. Many small newspapers in underdeveloped areas will for some time have to be serviced on a non-profit-making basis, possibly even at a temporary loss, if there is to be a real exchange of world news.

Only if the world news agencies show themselves fully prepared to undertake public services of this kind can they expect telecommunication administrations to make a comparable public service approach to problems inherent in the worldwide transmission of news to areas varying enormously in levels of economic development.

If newspapers and news agencies have public responsibilities in this respect, so also have telecommunication services—particularly since in many cases they are public monopolies. Press traffic should not, of course, be subsidized at the expense of other telecommunication users. It must pay its way. But it ought not necessarily to be expected to yield the same rate of profit as other traffic. Nor should it be automatically subject to regulations such as those applying to the joint use of leased lines, which may be applicable in other cases but are contrary to the public interest when applied to news.

This insistence on the special nature of press traffic is not new. It marked the early deliberations of the European members of the International Telegraph Union, when they considered the case for a press rate which would be half the ordinary rate. The French delegate made the same point at the Union's 1903 conference when he declared that a reduced rate for press messages was justified by the benefit it would bring in "the education of opinion and the diffusion of progressive ideas". The identical consideration has been a guiding principle in the operation of the penny press rate by the British Commonwealth telecommunications system.

It is against this background, and in the light of these long-established principles, that a new approach to the problem of telecommunications and the press should now be made.

As this survey has shown, the problem falls into two main parts.

First, there are problems which arise from the absence or comparatively obsolete nature of telecommunications in various parts of the world.

Secondly, there are problems which derive from high charges, restrictive regulations and anomalies within the telecommunication structures of more advanced countries.

Where barriers to a full exchange of world news arise from lack or obsolescence of equipment, a solution must largely depend upon developments in the telecommunication field as a whole.

As has been shown, there are still many areas where the absence of efficient internal telecommunication services is the major obstacle to a comprehensive system of news exchange. Within such areas, it is at present impossible for national news agencies to operate with complete effectiveness—and effective national news agencies are an essential link in any successful world system.

The part now being played by the ITU in the United Nations Technical Assistance Programme represents the beginning of a new approach to this vast problem. But funds available for the effort are severely limited. The amount allocated for technical assistance in the telecommunication field in 1952, for example, was no more than one per cent of the total special account, subject to a minimum of $200,000.

The ITU has undertaken to assist with surveys, with expert assistance in suggesting programmes of reorganization of telecommunication administrations in whole or in part, with the preparation of plans for new wire and radio networks and with the introduction of training for telecommunication staffs. It is arranging fellowships and scholarships, technical conferences, seminars and training centres.

There is great need for experts to give specialized assistance in all aspects of telecommunication development. It is therefore important that telecommunication administrations and private operating agencies should be prepared to make the necessary staff available, even at some temporary sacrifice to themselves.

Apart from what can be done through technical assistance, substantial capital sums will be required before internal telecommunication services in underdeveloped countries are brought up to the level necessary for completely satisfactory news exchange.

The extent to which telecommunications share in the capital resources available for social and economic reconstruction will materially depend on how far the public appreciates the urgency of the problem, and the essential part telecommunications can play in any effort to increase international understanding, develop adequate press services and reduce illiteracy. The press itself can do a great deal to arouse world public opinion to this need.

Even with full understanding of this requirement among governments and peoples, it is unlikely, however, that in the world's present economic circumstances, the large capital sums needed for development on the required scale will be obtainable except over a comparatively long period.

It is, therefore, all the more necessary to ensure maximum development of multiple address radio news services capable of reaching all parts of the world and of being received at a large number of scattered points. In many areas such international radio services, supplemented if possible by multiple address newscasts covering domestic news from a main national centre, could provide an overall distribution of news unattainable by any other means.

It is important that these newscasts should be made available to quite small newspapers in remote areas. Such publications have an important part to play in promoting international understanding, general education and knowledge of public affairs in areas which are at present economically and politically underdeveloped, but in which great advances and movements of opinion are possible in the near future. It is precisely these newspapers, the greater part of which lack any alternative means of securing information outside the restricted area of their circulation, that most need assistance to become informed and responsible vehicles of news and opinion capable of exercising an important social and educational influence in their communities.

Few such newspapers can afford to pay more than a very small amount for any news service provided. If their development as independent sources of news and comment is to be encouraged, they should not be forced to rely exclusively upon government-subsidized news services for information on matters outside their own locality. They therefore represent an important challenge to the big world news services and to the principle of a wide and free flow of independent basic news to which they express their fidelity.

In many areas, there cannot be great profit in supplying an outline service of basic world news which is all that such newspapers could handle. But a summarized report of the day's essential news, jointly sponsored by Reuters, the Associated Press, United Press, International News Service, Agence France-Presse and, if agreement could be reached, by Tass, and offered as an international public service at rates within the reach of small newspapers in under-developed areas without an independent external news service, could effect a revolutionary change of vast educational and social importance in the press of those countries.

Whether such a co-operative development is possible, or whether the problem can best be met by a greater extension of individual news agency services to underdeveloped areas, national and international telecommunication services must be ready to play their part. Radio reception by individual newspapers is an essential part of any such service, since its purpose is to overcome problems of internal news distribution which now prevent adequate coverage.

But, as we have seen, the administrations at present prepared to provide facilities for the transmission of multiple address newscasts are comparatively few. In addition, the direct reception of these newscasts by individual addressees is often prohibited, while elsewhere high reception fees put the service beyond the means of the smallest newspapers.

There is technically nothing to prevent direct "domestic" reception in even the smallest newspaper offices at quite low cost in equipment. Some of the existing Globereuter and other beam services are, in fact, already received in this way. But in many areas where such reception facilities are most needed, the regulations imposed by local administrations prevent their use.

Under the international agreement which was concluded at the Bermuda Tele-communications Conference of December 1945, and reaffirmed and extended at the London Conference of August 1949, the governments of the United States, United Kingdom and Canada agreed to "permit within their respective territories the private reception of such communications (multiple address press radio communications) either through the recipients' own radio receiving installation or through other private installations".

This principle, which is vitally important to the development of multiple address newscasts, is far from being generally accepted, however. Even at the Bermuda Conference, Australia, New Zealand, the Union of South Africa, India, and the United Kingdom (so far as its colonial territories were concerned) were prepared to arrange for reception only by telegraph administrations and insisted on retaining the power to exercise their discretion "as to the granting of permission to private recipients for the reception of such communications through their own installations or through other private installations".

Transmission and reception of multiple address newscasts are covered in the International Telegraph Regulations drawn up at the International Telephone and Telegraph Conference at Paris in 1949 (Chapter XXIV, Article 81). But administrations are left entirely free to organize or authorize services as they see fit. The administration in a country of reception may likewise decide whether authority should be given to the addressee to receive such communications. Under the same article the addressees "may be subjected by the administration of their country, apart from any charges levied for the estab-lishment and working of private receiving stations, to the payment of a receiver charge of which the amount and method of assessment shall be fixed by that administration".

These charges and payments vary considerably. So too does practice regard-ing authority for reception.

The attitude of administrations toward multiple address transmission and reception appears often to be governed by a determination to safeguard rates for the transmission of press material over point-to-point radio and cable circuits, and to limit competition with these more established methods of press trans-mission, even at the cost of crippling new methods. The refusal to allow direct reception is also undoubtedly influenced in some cases by the desire to exercise a form of censorship over news.

These limitations and restrictions must be ended if multiple address newscasting is to reach its full potential development; and, more particularly, if it is to become the means—as it alone can—of bringing a world news service to the populations of underdeveloped countries lacking the press or telecommunication structures within which other systems of news distribution can function.

The principle of free, unhampered access to news through direct private reception of newscasts by newspapers and agencies, which was accepted at Bermuda by the United States, the United Kingdom and Canada as applicable to their own territories, can alone ensure the abolition of all removable obstacles to worldwide news distribution. Its universal acceptance should be made a primary objective of international press and telecommunication policy.

Encouragement of new means to overcome physical barriers to the worldwide exchange of news is one of the most important tasks now remaining in the field of telecommunications. The sincerity of press and telecommunications organizations in the concern they express for the international public interest will be judged by their readiness to undertake this task.

But, as we have seen, obstacles to the free flow of news do not arise only from the absence of physical facilities. There are many hindrances to comprehensive world reporting even where adequate transmission services do exist.

Outstanding among them are the high cost of press messages from and to some parts of the world—notably the Middle East and parts of Latin America and Asia—and the serious anomalies in the rates. Many such anomalies result in substantial differences in the cost of sending messages between the same two points, according to whether they are going in one direction or the other.

Under Article 75 of the International Telegraph Regulations, 1949, it was agreed that "the terminal and transit rates applicable to ordinary press telegrams shall be those of ordinary private telegrams reduced by 50 per cent in the European system, and $66^2/_3$ per cent in other relations" and that the charge per word for an urgent press telegram should be the same as the rate per word for an ordinary private telegram over the same route.

If the distribution of news is to be regarded by purely commercial standards, such rates plainly represent a substantial concession to press users. This concession is, however, justified by the total volume of press traffic and by the fact that much of it is filed outside the peak hours for private and commercial operations.

But, as has been argued previously, the broad element of public interest involved in extending the flow of world news demands that other than commercial standards should be considered; and, particularly, that an attempt should be made to establish a structure of international press rates which will not impose on world reporting an arbitrary pattern unrelated to sound standards of newsworthiness.

The ideal way out of the present jungle of international press tariffs might well be that which has already been accepted as desirable and practical within the British Commonwealth telecommunication system. This is the largest and most varied international system in existence, since it embraces highly-developed countries with mature press and telecommunication services and colonial and non-self-governing territories where these services are still in their infancy.

A universal low press rate at the British Commonwealth level of a penny a word, or any such comparable level as might be determined by international agreement, would provide an immensely valuable stimulus to the world exchange of news.

Whether such a rate is economic, even within the Commonwealth system, is still far from clear. A much fuller analysis, utilizing all the evidence now available within the Commonwealth, is desirable.

At the Bermuda Telecommunications Conference, in which the United States, United Kingdom and other Commonwealth governments participated in 1945, the United Kingdom delegation submitted proposals for the general adoption of the penny rate. This, however, was rejected by the United States delegation on the grounds that an independent press telecommunication service such as Press Wireless (U.S.A.) could not operate on it and that such a rate was bound to imply a subsidy of press traffic by other business.

The great increase in Commonwealth press traffic resulting from the penny rate and (leaving aside the public interest involved in such an increase) the consequent expansion in revenue from press traffic would suggest, however, that this is not necessarily the case.

When the Seventh Imperial Press Conference met in Canada in 1950, W. A. Cole, editor of Reuters, urged that if the penny press rate should be threatened, the Commonwealth Press Union, representing the newspapers of some 14 countries at varying stages of press development, should seek powers to operate its own press telecommunication service. He voiced the opinion, based on wide experience in international press communications, that a telecommunication co-operative, handling only press traffic, could maintain or even reduce the penny rate.

Apart, however, from the ultimate objective of a universal low press rate, the wide variations and serious anomalies in the present rate structure clearly call for urgent review. It would therefore be of immense advantage if the ITU would consider the appointment of a joint consultative committee, representing the ITU itself, international press organizations, the major world agencies and national agencies or newspapers of countries with press structures at varying stages of development, to examine the whole question of press telecommunication facilities in the light of the international public interest.

Such a consultative committee could have a status comparable to that of the ITU's existing consultative committees on radio-telephone and telegraph questions. Its first task would be to make a careful review of methods of handling press traffic with a view to proposing, for consideration by the ITU's appropriate Administrative Conferences, such as the International Telephone and Telegraph Conference, means to reduce the present wide gap between the highest and lowest rates in the international system, and to establish a common charge for transmissions between the same points in either direction.

As a long-term task, the committee could invite the British Commonwealth Telecommunications Board to make available to it all data relevant to a full examination of the effect of the penny press rate (a) on the volume of press traffic and (b) on the cost of handling such traffic. The consultative committee could then investigate the extent to which the British Commonwealth experience might be generally applicable, and how far it might provide a basis for proposals for a universal international press rate to be considered by the ITU's Administrative Conferences and its Plenipotentiary Conference.

As we have seen, however, wide disparities in press cable and radio-telegraph rates are not the only obstacles to the wider exchange of news, although they are among the most serious. There is also need for an urgent review of rates and procedures concerning leased channels, particularly in relation to (a) the considerable differences in rates between adjacent countries within the European system; (b) anomalies arising from the practice of fixing different rates for identical channels, according to whether they are leased for internal transmissions or as part of an international link; and (c) obstacles to the co-operative use of leased channels by news agencies and newspapers, either as part of a permanent news exchange service, or to assist the coverage of internationally important events.

Finally, there is need for re-examination of present facilities for international telephoto services, to see how far technical advances or administrative reforms can bring about a closer equalization of rates and an extension of facilities.

It is greatly to be hoped that the telecommunication administrations primarily concerned in leasing international teleprinter networks in Europe will themselves give close consideration to problems arising under these heads. They should do this, if possible, in conjunction with the committee appointed by the European Technical Conference of Press Agencies to examine this matter.

Here, also, an ITU consultative committee, with the status and duties suggested, could be of immense assistance. The appointment of such a committee, bringing together press and telecommunication interests in a permanent endeavour to improve in every possible way the means by which news may be exchanged among peoples, would provide concrete evidence of the readiness of both groups to serve the public interest in a vitally important field.

From such an effort could flow practical results of great value. . It is therefore to be hoped that the ITU, with its long and successful record of international co-operation, will be prepared to take the necessary initiative.

Advances in press and telecommunications during the past century have been of immense importance in the history of civilization. The part which these two groups can play, now and in the future, in advancing knowledge and promoting international understanding is so great that continuing machinery for joint co-operation in overcoming common problems is urgently required. These problems are not theirs alone. They constitute an essential part of human progress and as such concern all men and all nations. Upon their solution depends much that is essential to the happiness and well-being of international society.

recommendations

In the light of the facts examined in this survey, I would submit the following suggestions.

1. In view of the urgent need to make available to the peoples of underdeveloped areas a basic and independent service of essential world news, the measures outlined below might be given immediate consideration by the world news agencies, by telecommunication administrations and private operators, and by the International Telecommunication Union :

    (a) agreement by the major world news services to examine the possibility of transmitting to areas where basic news services are restricted or non-existent, either separate or pooled multiple address broadcast summaries of essential news on a non-profit-making basis for direct receipt by small and scattered newspapers. Such newspapers would be required, during the initial period of their development, to pay a subscription determined solely on a cost basis and levied according to circulation ;

    (b) acceptance by telecommunication administrations of a share in providing this public service by making available transmitters at the lowest economic cost, calculated on a non-profit-making basis, for beaming such co-operative multiple address newscasts to areas lacking news facilities, and by permitting the direct receipt of such newscasts by the newspapers concerned.

2. In view of the need to extend multiple address distribution of national and world news in many areas which, while not within the above category, can be serviced by news agencies according to their customary practice and within the framework of existing beam services, members of the ITU might be invited to consider action to:
   (a) extend facilities for the transmission of such services;
   (b) reduce present reception charges to levels consistent with actual reception costs;
   (c) promote general international agreement, on the lines of the Bermuda Agreement between the United States, the United Kingdom and Canada which permits private reception of multiple address press radio communications within their territories, either through the recipient's own radio receiving installation or through other private installations.

3. Members of the ITU might be invited to consider action to secure maximum endorsement of the United States declaration, published as an appendix to the International Telegraph Regulations, urging telecommunication administrations not to use the authority given them by the regulations to restrict the reception of press radio communications.

4. In view of the urgent need to reduce barriers which hinder the free flow of news and the proper reporting of news and opinion and which result from the high cost of press cables and radiograms to and from important areas (notably the Middle East, South America and parts of Asia); and in consequence of the many striking anomalies in the international press rates structure members of the ITU, might be invited urgently to consider the desirability of establishing a permanent consultative committee to examine and report on press telecommunications.

   This consultative committee should enjoy the same status as the existing consultative committees of the ITU. It should include, in addition to such ITU membership as was felt desirable, representatives of world news agencies and the newspapers of countries with both highly developed and less developed press structures. The committee should report direct to the appropriate Administrative Conferences of the ITU and, where necessary, to the Plenipotentiary Conference.

5. The consultative committee should be given the following specific tasks in addition to its continuing responsibilities to keep under constant review the best means of securing the maximum telecommunication facilities for the exchange of world news.

   It should be required to make an immediate review of methods of handling press traffic, and of the costs which enter into the handling of such traffic, with a view to:
   (a) advising on the best means to reduce the present wide gap between the highest and lowest press rates in the international system;
   (b) establishing the same charges for transmissions between the same

87

points in either direction. Special attention would be paid to available evidence as to the effect of rate reductions in increasing the volume of press traffic and thus the total income from this source.

6. As part of its efforts to obtain reductions in international press rates, the consultative committee might seek the co-operation of the British Commonwealth Telecommunications Board in making a thorough investigation of the effect of the Commonwealth penny rate:
   (a) on the volume of Commonwealth press traffic;
   (b) on handling costs.

   The purpose of this investigation would be to determine the true economic basis of this rate, and to discover how far the experience of the Commonwealth Telecommunications Board is generally applicable and capable of providing a basis for proposals for a universal cheap press rate, to be considered by the appropriate Administrative Conference of the ITU.

7. In view of the importance of efficient teleprinter networks for the exchange of news in well-developed areas, the European telecommunication administrations, in association with the consultative committee, might be invited to examine:
   (a) the causes of the great variations between the rates charged for leased lines in adjacent countries within the European system (or elsewhere), with a view to securing greater uniformity;
   (b) the anomalies arising from the practice of fixing different rates for identical channels according to whether they are used for national or international news distribution, with a view to removing such anomalies. In this examination, the co-operation of the Special Committee of the European Technical Conference of Press Agencies could be invited.

8. In view of the importance of promoting maximum co-operation and exchange between world and national agencies and of ensuring maximum coverage of major international events, telecommunication administrations might be invited, in association with the proposed consultative committee, to consider revision of present regulations relating to shared lines. Such regulations hinder the co-operative organization of permanent telecommunication networks by national and world news agencies and joint participation by agencies or newspapers in leasing special channels for economic coverage of important events.

9. The consultative committee might be invited to examine obstacles hindering direct news reporting by radio organizations, which are frequently required to arrange advance booking of channels for unnecessarily long periods and also in some cases for additional test periods. The committee might examine the evidence as to the effect of these practices in unnecessarily curtailing the channels available for ordinary press traffic.

10. In view of the increasing value of pictures as a means of international mass communication and the fact that many photographs have a universality of appeal beyond that of the written word, the consultative committee might be invited to investigate and report on the present position of international telephoto channels with a view to:
    (a) improving facilities where these are inadequate;
    (b) securing equalization of rates for telephoto transmissions.

The American Press and International Communications, Louis Caldwell, Chicago 1945.

Annual Report of the Commonwealth Telecommunications Board, London 1952.

A.P.: The Story of News, Oliver Gramling, New York 1940.

Barriers Down, Kent Cooper, New York 1942.

Das Bild als Nachricht: Nachrichtenwert und Technik des Bildes, Willy Stiewe, Berlin–1933.

Communication Networks and Lines, Walter J. Creamer, New York 1951.

Communications in Modern Society, Wilbur Schramm (editor), Urbana, Illinois 1948.

Communications Research, Paul F. Lazarsfeld and F. N. Stanton, New York 1941-43.

Deutsche Presse Agentur (DPA), DPA, Hamburg 1950.

Facsimile, Lee Hills and T. J. Sullivan, New York 1949.

The First Freedom, Morris L. Ernst, New York 1946.

A Free and Responsible Press, Commission on Freedom of the Press, Chicago 1947.

Freedom of Information, Herbert Brucker, New York 1949.

Freedom of Information (annotated bibliography), Library of Congress, Washington D.C. 1952.

Freedom of Information, United Nations, New York 1950.

Freedom of the Press—An International Issue, Samuel de Palma, Washington D.C. 1950.

Government and Mass Communications, Zachariah Chaffee, Chicago 1947.

History of "The Times", Vol. II (1841-1884), The Times, London 1951.

Improvement of Information, International Press Institute, Zürich 1952.

International News and the Press (annotated bibliography), Ralph O. Nafziger, New York 1940.

International Telecommunication Convention 1947: Telephone and Telegraph Regulations and Final Protocols (Paris Revision 1949), International Telecommunication Union, Geneva.

International Telecommunications, Osborne Mance, London 1944.

Legislation for Press, Film and Radio, F. Terrou and L. Solal (Unesco), Paris 1951.

La Liberté de la Presse, Jacques Bourquin, Paris 1950.

A Life in Reuters, Roderick Jones, London 1951.

List of International Telegraph Channels, International Telecommunication Union, Geneva 1952.

List of Point-to-Point Radiocommunication Channels, International Telecommunication Union, Geneva 1952.

List of Submarine Cables, International Telecommunication Union, Geneva 1951.

Mass Communications, Wilbur Schramm (editor), Urbana, Illinois 1949.

Die Nachricht im Weltverkehr, Wilhelm Schwedler, Berlin 1932.

Der Nachrichtenschnellverkehr im Dienste von Presse und Wirtschaft, Friedrich Winkin, Leipzig 1934.

Die Nachrichtenübermittlung im Wandel der Zeiten: von Feuerzeichen zur Funkwelle, Otto Lencke, Berlin 1941.

Newspapers, David Keir, London 1948.

Peoples Speaking to Peoples, Commission on Freedom of the Press, Chicago 1946.

The Press and Society, G. L. Bird and F. E. Merwin, New York 1951.

Press, Radio, Film, Vols. I-V, Unesco, Paris 1947-52.

Print, Radio and Film in a Democracy, Douglas Waples, Chicago 1942.

Problèmes et Techniques de Presse, Fondation Nationale des Sciences Politiques, Paris 1950.

Public Opinion in Soviet Russia, Alexander Inkeles, Cambridge, Massachusetts 1950.

Radio, Television and Society, Carl A. Siepmann, New York 1950.

Report of the Commonwealth Press Conference, 1950, Commonwealth Press Union, London 1951.

Report of the Imperial Press Conference, 1939, Empire Press Union, London 1940.

Report of the Royal Commission on the Press, 1947-1949, H.M. Stationery Office, London 1949.

Reuters Century, 1851-1951, Graham Storey, London 1951.

Telecommunications: A Program for Progress, President's Communications Policy Board, Washington D.C. 1951.

"Die Telegraphenagentur der Sowjetunion," J. Doletzky, *Zeitungswissenschaft*, 2 Jahrgang, Berlin 1926.

News Agencies: Their Structure and Operation, Unesco, Paris 1953.

World Communications: Press, Radio, Film, Television, Unesco, Paris 1951.

90

91

Spain, 24, 29, 65, 66, 71.
Submarine Company, U.K., 21
Sudan, 59.
Sundell, Colonel Olof, 66.
Sweden, 29, 48, 65, 66, 71, 72, 76.
Switzerland, 20, 24, 34, 59, 65, 66, 71, 76.
Syria, 59, 60.

Taganyika, 29.
Tanjug (Yugoslav news agency), 65.
Tass, 39-42, 82.
Taylor, T. E., 20.
Telegraph and telephone services, 17-23, 27-32, 47-49, 58, 66, 69, 70.
Teleprinter line rental rates, 66-68, 85, 88.
Teleprinter networks, 39-41, 45-50, 58, 64, 65, 68, 69, 73, 85, 88.
Thailand, 47, 50 ; press, 50.
The Times, London, 12, 17, 19-21.
Tidningarnas Telegrambyra (TT— Swedish news agency), 65, 66.
Transocean news agency, Germany, 43.
Tripolitania, 29.
Trollope, T. A., 17.
Tropical Radio Telegraph Company, 70.
Tunis, 61.
Turkey, 23, 24, 59, 60, 71.

Ultrafax, 74.
Underdeveloped areas, 25-27, 31, 36, 48, 52, 75, 80-84.
Union of Soviet Socialist Republics, 23, 24, 29, 32, 39-42, 48, 76.
United Kingdom, 17, 20-23, 28-32, 34, 39, 41, 42, 47, 48, 59, 60, 65, 66, 71, 76, 83, 84, 87 ; press, 20, 22, 39, 47, 48, 60 ; Post Office, 20, 23, 28, 30, 31, 41, 44, 45, 47, 53, 56, 67, 70.

United Nations, 33, 52.
United Nations Educational, Scientific and Cultural Organization (Unesco), 17, 41, 42, 47.
United Nations Technical Assistance Programme, 36, 73, 81.
United Press Associations (UP), 39-45, 47-49, 51, 71, 82.
United States of America, 24, 26, 28-32, 39-42, 45, 47, 48, 56, 65-67, 70-72, 74-76, 83, 84, 87 ; press, 24, 40, 41, 47, 63 ; news agencies, 30, 65, 70, 71 (see also AP, UP, INS).
Universal Declaration of Human Rights, 26, 27, 47.
Uruguay, 29.

Venezuela, 32, 61.
Viet-Nam, 71.

West Indies, 28, 29, 61.
Western Associated Press, Chicago, U.S.A., 23.
Western Union Telegraph Company, 28, 30, 31, 47, 75.
White, Llewellyn, 57, 63, 71.
Wilshaw, Sir Edward, 53, 54.
Wolff (German news agency), 18, 19, 23, 24.
World events which stimulated tele-communication development :
    American Civil War, 21-23.
    Crimean War, 20, 21.
    Franco-Prussian War, 21.
    Indian Mutiny, 21, 22.
    World War I, 24, 43.
    World War II, 45.
World News Agencies, 13, 18, 19. 23-26, 39-46, 63, 81, 86.

Yugoslavia, 65.

INTERNATIONAL PROPAGANDA AND COMMUNICATIONS

An Arno Press Collection

Bruntz, George G. **Allied Propaganda and the Collapse of the German Empire in 1918.** 1938

Childs, Harwood Lawrence, editor. **Propaganda and Dictatorship:** A Collection of Papers. 1936

Childs, Harwood L[awrence] and John B[oardman] Whitton, editors. **Propaganda By Short Wave** *including* C[harles] A. Rigby's **The War on the Short Waves.** 1942/1944

Codding, George Arthur, Jr. **The International Telecommunication Union:** An Experiment in International Cooperation. 1952

Creel, George. **How We Advertised America.** 1920

Desmond, Robert W. **The Press and World Affairs.** 1937

Farago, Ladislas, editor. **German Psychological Warfare.** 1942

Hadamovsky, Eugen. **Propaganda and National Power.** 1954

Huth, Arno. **La Radiodiffusion Puissance Mondiale.** 1937

International Propaganda/Communications: Selections from *The Public Opinion Quarterly*, 1943/1952/1956. 1972

International Press Institute Surveys, Nos. 1-6. 1952-1962

International Press Institute. **The Flow of News.** 1953

Lavine, Harold and James Wechsler. **War Propaganda and the United States.** 1940

Lerner, Daniel, editor. **Propaganda in War and Crisis.** 1951

Linebarger, Paul M. A. **Psychological Warfare.** 1954

Lockhart, Sir R[obert] H. Bruce. **Comes the Reckoning.** 1947

Macmahon, Arthur W. **Memorandum on the Postwar International Information Program of the United States.** 1945

de Mendelssohn, Peter. **Japan's Political Warfare.** 1944

Nafziger, Ralph O., compiler. **International News and the Press:** An Annotated Bibliography. 1940

Read, James Morgan. **Atrocity Propaganda, 1914-1919.** 1941

Riegel, O[scar] W. **Mobilizing for Chaos:** The Story of the New Propaganda. 1934

Rogerson, Sidney. **Propaganda in the Next War.** 1938

Summers, Robert E., editor. **America's Weapons of Psychological Warfare.** 1951

Terrou, Fernand and Lucien Solal. **Legislation for Press, Film and Radio:** Comparative Study of the Main Types of Regulations Governing the Information Media. 1951

Thomson, Charles A. H. **Overseas Information Service of the United States Government.** 1948

Tribolet, Leslie Bennett. **The International Aspects of Electrical Communications in the Pacific Area.** 1929

Unesco. **Press Film Radio,** Volumes I-V *including* Supplements. 1947-1951. 3 volumes.

Unesco. **Television:** A World Survey *including* Supplement. 1953/1955

White, Llewellyn and Robert D. Leigh. **Peoples Speaking to Peoples:** A Report on International Mass Communication from The Commission on Freedom of the Press. 1946

Williams, Francis. **Transmitting World News.** 1953

Wright, Quincy, editor. **Public Opinion and World-Politics.** 1933